An OPUS book

# THE VOICE OF THE PAST

OPUS General Editors

Keith Thomas
Alan Ryan
Walter Bodmer

OPUS books provide concise, original, and authoritative introductions to a wide range of subjects in the humanities and sciences. They are written by experts for the general reader as well as for students.

# The
# Voice of the Past

## Oral History

**Second Edition**

PAUL THOMPSON

Oxford   New York

OXFORD UNIVERSITY PRESS

1988

Oxford University Press, Walton Street, Oxford OX2 6DP

Oxford New York Toronto
Delhi Bombay Calcutta Madras Karachi
Petaling Jaya Singapore Hong Kong Tokyo
Nairobi Dar es Salaam Cape Town
Melbourne Auckland

and associated companies in
Beirut Berlin Ibadan Nicosia

Oxford is a trade mark of Oxford University Press

First published 1978 as an Oxford University Press paperback
and simultaneously in a hardback edition
Paperback reprinted 1982, 1984
Second edition 1988

British Library Cataloguing in Publication Data

Thompson, Paul, 1935–
The voice of the past : oral history.
——2nd ed.
1. Oral history
I. Title
907   D16.14

ISBN 0–19–219230–2
ISBN 0–19–289216–9 Pbk

Library of Congress Cataloging in Publication Data

Thompson, Paul Richard, 1935–
The voice of the past. (OPUS)
Bibliography: p.      Includes index.
1. Oral history.   2. History—Methodology.
I. Title.   II. Series.
D16.14.T48   1988   907'.2   87–21984

ISBN 0–19–219230–2
ISBN 0–19–289216–9 (pbk.)

Printed in Great Britain by
Biddles Ltd.
Guildford and King's Lynn

*To Natasha*

# Preface to the First Edition

THIS is a book about both the method, and the meaning, of history. It is, first of all, an introduction to the use of oral sources by the historian. But the very use of these sources raises fundamental issues, and I have decided to take these at the beginning, moving step by step towards the more practical later chapters. At the same time, I have tried to write with many different types of reader in mind. Some may be more immediately concerned with how to design a project, and to collect and evaluate interview material. They will find practical advice in Chapters 6 (Projects), 7 (The Interview), and 8 (Storing and Sifting). There would be, indeed, good sense in starting from field-work. The practical experience of oral history will itself lead on to deeper questions about the nature of history.

These concern, first, the character of evidence. How reliable is oral evidence? How does it compare with the modern historian's more familiar documentary sources? These critical and immediate questions are confronted in Chapter 4 (Evidence). But they are better understood when placed within the wider context of the development of historical writing. Chapter 3 (The Achievement of Oral History) provides an assessment of recent writing and the contribution which oral evidence has made in providing new perspectives and opening up fresh fields of inquiry. Chapter 2 (Historians and Oral History) pursues the question back into the past of history itself, exploring the changing approach of historians to evidence, from the original primacy of oral tradition to the eras of the written document and the tape recorder.

But inevitably this leads to a second set of questions concerning the social function of history. Indeed, it became clear in writing Chapter 2 that the evolution of scholarly techniques could only be convincingly explained in such a social context. And the problems in selecting and evaluating oral evidence had already pointed in the same direction. How do we choose who to listen to? History survives as social activity only because it has a meaning for people today. The voice of the past matters to the present. But whose voice—or voices—are to be heard?

Thus, while method and meaning can be treated as independent themes, they are at bottom inseparable. The choice of evidence must reflect the role of history in the community. This is in part a political question, on which historians can only reach their own position independently. Consequently, although even here most of the argument is straightforwardly human rather than political, Chapter 1 (History and the Community) is written from a socialist perspective. And I myself believe that the richest possibilities for oral history lie within the development of a more socially conscious and democratic history. Of course, a telling case could equally be made, from a conservative position, for the use of oral history in perserving the full richness and value of tradition. The merit of oral history is not that it entails this or that political stance, but that it leads historians to an awareness that their activity is inevitably pursued within a social context and with political implications.

This, then, is a practical book about how oral sources can be collected and used by historians. But it is equally intended to provoke historians to ask themselves what they are doing, and why. On whose authority is their reconstruction of the past based? For whom is it intended? In short, *whose* is *The Voice of the Past*?

I have been fortunate in writing in being able to depend on the help of many friends and colleagues, especially in the Oral History Society and at the University of Essex. It is impossible to acknowledge more than a very few separately but I should like to thank them all. In ten years of research activity and student projects, and a widening circle of discussions and conferences, experiments, mistakes, and successes have built up a widely shared collective experience. It is on this that the book rests. Above all, it draws on the joint work, in research and then in graduate teaching, through which Thea Vigne and I came upon and ourselves explored the possibilities of oral evidence in social history. I owe an immeasurable debt to her. I should also like to thank again those others who are mentioned in the preface to the first-fruit of that research, *The Edwardians*, and particularly

George Ewart Evans and Mary Girling. And for this present text, for specific contributions to it and for comments on earlier drafts, I am especially grateful to Keith Thomas, Geoffrey Hawthorn, Bill Williams, Colin Bundy, Trevor Lummis, Roy Hay, Michael Winstanley, Gina Harkell, Joanna Bornat, Alun Howkins, Eve Hostettler, Natasha Burchardt, and Raphael Samuel.

# Preface to the Second Edition

IN the ten years since I first wrote this book much has happened. The work which was then starting has resulted in some outstanding published history. We have moved forward in our understanding of the complexity of the memory process and the interpretation of oral sources. We have watched a vigorous spread of local community projects, and the rise of new movements in adult literacy, drama, and reminiscence therapy. We have learnt more about the past of oral history. We have developed firmer links with life story sociology, and we have joined together to form an international community of oral historians.

All these developments are reflected in this new edition. In particular, I have expanded the first three chapters on history and the community, historians and oral history, and the achievement of oral history; I have introduced a new discussion of 'subjectivity', psychoanalysis, and memory as therapy in a rewritten chapter on evidence (4) and a new chapter on memory and the self (5); and I have rewritten and expanded the final chapter on interpretation (9).

I wish I could thank all those who have helped me over the years with the rethinking and experience which is reflected in this revised edition; but more than ever, they are too many to mention individually. To those British friends and colleagues to whom I continue to owe so much I would now want to add many from other countries, and especially many generous hosts on journeys in Scandinavia, Poland, France and Italy, Belgium and Spain, North America, and China. Let me simply salute Ron Grele for his crucial role in creating an international forum for oral history; and indicate the very special personal debt which I owe to Daniel Bertaux, Isabelle Bertaux-Wiame, Luisa Passerini, and our other friends of Turin, with whom we have worked on common research projects over these years.

# Contents

# 1

# History and the Community

ALL history depends ultimately upon its social purpose. This is why in the past it has been handed down by oral tradition and written chronicle, and why today professional historians are supported from public funds, children are taught history in schools, amateur history societies blossom, and popular history books rank among the strongest bestsellers. Sometimes the social purpose of history is obscure. There are academics who pursue fact-finding research on remote problems, avoiding any entanglement with wider interpretations or contemporary issues, insisting only on the pursuit of knowledge for its own sake. They have one thing in common with the bland contemporary tourism which exploits the past as if it were another foreign country to escape to: a heritage of buildings and landscape so lovingly cared for that it is almost inhumanly comfortable, purged of social suffering, cruelty, and conflict to the point that a slavery plantation becomes a positive pleasure. Both look to their incomes free from interference, and in return stir no challenge to the social system. At the other extreme the social purpose of history can be quite blatant: used to provide justification for war and conquest, territorial seizure, revolution and counter-revolution, the rule of one class or race over another. Where no history is readily at hand, it will be created. South Africa's white rulers divide their urban blacks between tribes and 'homelands'; Welsh nationalists gather at bardic eisteddfods; the Chinese of the cultural revolution were urged to construct the new 'four histories' of grass-roots struggle; radical feminists looked to the history of wet-nursing in their search for mothers without maternal instinct. Between these two extremes are many other purposes, more or less obvious. For politicians the past is a quarry for supportive symbols: imperial victories, martyrs, Victorian values, hunger marches. And almost equally telling are the gaps in the public presentation of history: the silences in Russia on Trotsky, in West Germany on the Nazi era, in France on the Algerian war.

Through history ordinary people seek to understand the upheavals and changes which they experience in their own lives: wars, social transformations like the changing position of youth, technological changes like the end of steam power, or personal migration to a new community. Family history especially can give an individual a strong sense of a much longer personal lifespan, which will even survive their own death. Through local history a village or town seeks meaning for its own changing character and newcomers can gain a sense of roots in personal historical knowledge. Through political and social history taught in schools children are helped to understand, and accept, how the political and social system under which they live came about, and how force and conflict have played, and continue to play, their part in that evolution.

The challenge of oral history lies partly in relation to this essential social purpose of history. This is a major reason why it has so excited some historians, and so frightened others. In fact, fear of oral history as such is groundless. We shall see later that the use of interviews as a source by professional historians is long-standing and perfectly compatible with scholarly standards. American experience shows clearly enough that the oral history method can be regularly used in a socially and politically conservative manner; or indeed pushed as far as sympathy with Fascism in John Toland's portrait of *Adolf Hitler* (New York, 1976).

Oral history is not necessarily an instrument for change; it depends upon the spirit in which it is used. Nevertheless, oral history certainly can be a means for transforming both the content and the purpose of history. It can be used to change the focus of history itself, and open up new areas of inquiry; it can break down barriers between teachers and students, between generations, between educational institutions and the world outside; and in the writing of history—whether in books, or museums, or radio and film—it can give back to the people who made and experienced history, through their own words, a central place.

Until the present century, the focus of history was essentially political: a documentation of the struggle for power, in which the lives of ordinary people, or the workings of the economy or religion, were given little attention except in times of crisis such as the Reformation, the English Civil War, or the French Revolution. Historical time was divided up by reigns and dynasties. Even local history was concerned with the administration of the hundred and parish rather than the day-to-day life of the community and the street. This was partly because historians, who themselves then belonged to the administering and governing classes, thought that this was what mattered most. They had developed no interest in the point of view of the labourer, unless he was specifically troublesome; nor—being men—would they have wished to inquire into the changing life experiences of women. But even if they had wished to write a different kind of history, it would have been far from easy, for the raw material from which history was written, the documents, had been kept or destroyed by people with the same priorities. The more personal, local, and unofficial a document, the less likely it was to survive. The very power structure worked as a great recording machine shaping the past in its own image.

This has remained true even after the establishment of local record offices. Registers of births and marriages, minutes of councils and the administration of poor relief and welfare, national and local newspapers, schoolteachers' log books—legal records of all kinds are kept in quantity; very often there are also church archives and accounts and other books from large private firms and landed estates, and even private correspondence from the ruling landowner class. But of the innumerable postcards, letters, diaries, and ephemera of working-class men and women, or the papers of small businesses like corner shops or hill farmers, for example, very little has been preserved anywhere.

Consequently, even as the scope of history has widened, the original political and administrative focus has remained. Where ordinary people have been brought in, it has been generally as statistical aggregates derived from some earlier administrative investigation. Thus economic history is constructed around three types of source: aggregate rates of wages, prices, and unemployment; national and international political interventions into the economy and the information which arises from these; and studies of particular trades and industries, depending on the bigger and more successful firms for records of individual enterprises. Similarly, labour history for long consisted of studies on the one hand of the relationship between the working classes and the state in general, and on the other of particular but essentially institutional accounts of trade unions and working-class political organizations; and, inevitably, it is the larger and more successful organizations which normally leave records or commission their own histories. Social history has remained especially concerned with legislative and administrative developments like the rise of the welfare state; or with aggregate data such as population size, birth rates, age at marriage, household and family structure. And among more recent historical specialisms, demography has been almost exclusively concerned with aggregates; the history of the family, despite some ambitious but ill-judged attempts to break through to a history of emotion and feeling, has tended to follow the lines of conventional social history; while at least until quite recently women's history has to a remarkable extent focused on the political struggle for civil equality, and above all for the vote.

There are, of course, important exceptions in each of these fields, which show that different approaches are possible even with the existing sources. And there is a remarkable amount of unexploited personal and ordinary information even in official records—such as court documents—which can be used in new ways. The continuing pattern of historical writing probably reflects the priorities of the majority of the profession—even if no longer of the ruling class itself—in an age of bureaucracy, state power, science, and statistics. Nevertheless, it remains true that to write any other kind of history from documentary sources remains a very difficult task, requiring special ingenuity. It is indicative of the situation that E. P. Thompson's *The Making of the English Working Class* (1963) and James Hinton's *The First Shop Stewards' Movement* (1973) each depended to a large extent on reports by paid government informers, in the early nineteenth century and First World War respectively. When socialist historians are reduced to writing history from the records of government spies, the constraints imposed are clearly extreme. We cannot, alas, interview tombstones, but at least for the First

World War period and back into the late nineteenth century, the use of oral history immediately provides a rich and varied source for the creative historian.

In the most general sense, once the life experience of people of all kinds can be used as its raw material, a new dimension is given to history. Oral history provides a source quite similar in character to published autobiography, but much wider in scope. The overwhelming majority of published autobiographies are from a restricted group of political, social, and intellectual leaders, and even when the historian is lucky enough to find an autobiography from the particular place, time, and social group which he happens to need, it may well give little or no attention to the point at issue. Oral historians, by contrast, may choose precisely whom to interview and what to ask about. The interview will provide, too, a means of discovering written documents and photographs which would not have otherwise been traced. The confines of the scholar's world are no longer the well-thumbed volumes of the old catalogue. Oral historians can think now as if they themselves were publishers: imagine what evidence is needed, seek it out and capture it.

For most existing kinds of history, probably the critical effect of this new approach is to allow evidence from a new direction. The historian of working-class politics can juxtapose the statements of the government or the trade union headquarters with the voice of the rank and file—both apathetic and militant. There can be no doubt that this should make for a more realistic reconstruction of the past. Reality is complex and many-sided; and it is a primary merit of oral history that to a much greater extent than most sources it allows the original multiplicity of standpoints to be recreated. But this advantage is important not just for the writing of history. Most historians make implicit or explicit judgements—quite properly, since the social purpose of history demands an understanding of the past which relates directly or indirectly to the present. Modern professional historians are less open with their social message than Macaulay or Marx, since scholarly standards are seen to conflict with declared bias. But the social message is usually present, however obscured. It is quite easy for a historian to give most of his attention and quotations to those social leaders whom he admires, without giving any direct opinion of his own. Since the nature of most existing records is to reflect the standpoint of authority, it is not surprising that the judgement of history has more often than not vindicated the wisdom of the powers that be. Oral history by contrast makes a much fairer trial possible: witnesses can now also be called from the under-classes, the unprivileged, and the defeated. It provides a more realistic and fair reconstruction of the past, a challenge to the established account. In so doing, oral history has radical implication for the social message of history as a whole.

At the same time oral history implies for most kinds of history some shift of focus. Thus the educational historian becomes concerned with the experiences of children and students as well as the problems of teachers and administrators. The military and naval historian can look beyond command level strategy and equipment to the conditions, recreations, and morale of other ranks and the lower deck. The social historian can turn from bureaucrats and politicians to poverty itself, and learn how the poor saw the relieving officer and how they survived his refusals. The political historian can approach the voter at home and at work; and can hope to understand even the working-class conservative, who produced no newspapers or organizations for investigation. The economist can watch both employer and worker as social beings and at their ordinary work, and so come closer to understanding the typical economic process, and its successes and contradictions.

In some fields, oral history can result not merely in a shift in focus, but also in the opening up of important new areas of inquiry. Labour historians, for example, are enabled for the first time to undertake effective studies of the ill-unionized majority of male workers, of women workers, and of the normal experience of work and its impact on the family and the community. They are no longer . confined to those trades which were unionized, or those which gained contemporary publicity and investigation because of strikes or extreme poverty. Urban historians similarly can turn from well-explored problem areas like the slums to look at other typical forms of urban social life: the small industrial or market town, for example, or the middle-class surburb, constructing the local patterns of social distinctions, mutual help between neighbours and kin, leisure and work. They can even approach from the inside the history of immigrant groups—a kind of history which is certain to become more important in Britain, and is mainly documented only from outside as a social problem. These opportunities—and many others—are shared by social historians: the study of working-class leisure and culture, for example; or of crime from the point of view of the ordinary, often undetected and socially semi-tolerated poacher, shoplifter, or work-pilferer.

Perhaps the most striking feature of all, however, is the transforming impact of oral history upon the history of the family. Without its evidence, the historian can discover very little indeed about either the ordinary family's contacts with neighbours and kin, or its internal relationships. The roles of husband and wife, the upbringing of girls and boys, emotional and material conflicts and dependence, the struggle of youth for independence, courtship, sexual behaviour within and outside marriage, contraception and abortion—all these were effectively secret areas. The only clues were to be gleaned from aggregate statistics, and from a few—usually partial—observers. The historical paucity which results is well summed up in Michael Anderson's brilliant, speculative, but abstract study of *Family Structure in Nineteenth-Century Lancashire* (1971): a lop-sided, empty frame. With the use of interviewing, it is now possible to develop a much fuller history of the family over the last ninety years, and to establish its main patterns and changes over time, and from place to place, during the life cycle and between the sexes. The history of childhood as a whole becomes practicable for the first time. And given the dominance of the family through housework, domestic service, and motherhood in the lives of most women, an almost equivalent broadening of scope is brought to the history of women.

In all these fields of history, by introducing new evidence from the underside, by shifting the focus and opening new areas of inquiry, by challenging some of the assumptions and accepted judgements of historians, by bringing recognition to substantial groups of people who had been ignored, a cumulative process of transformation is set in motion. The scope of historical writing itself is enlarged and enriched; and at the same time its social message changes. History becomes, to put it simply, more democratic. The chronicle of kings has taken into its concern the life experience of ordinary people. But there is another dimension to this change, of equal importance. The process of writing history changes along with the content. The use of oral evidence breaks through the barriers between the chroniclers and their audience; between the educational institution and the outside world.

This change springs from the essentially creative and co-operative nature of the oral history method. Of course oral evidence once recorded can be used by lone scholars in libraries just like any other type of documentary source. But to be content with this is to lose a key advantage of the method: its flexibility, the ability to pin down evidence just where it is needed. Once historians start to interview they find themselves inevitably working with others—at the least, with their informants. And to be a successful interviewer a new set of skills is needed, including an understanding of human relationships. Some people can find these skills almost immediately, others need to learn them; but in contrast to the cumulative process of learning and amassing information which gives such advantage in documentary analysis and interpretation to the professional historian well on in life, it is possible to learn quite quickly to become an effective interviewer. Hence historians as field-workers, while in important respects retaining the advantages of professional knowledge, also find themselves off their desks, sharing experience on a human level.

Because of these characteristics, oral history is peculiarly suited to project work—both for groups and for individual student enterprises: in schools, universities, colleges, adult education, or community centres. It can be carried out anywhere. In any part of the country there is an abundance of topics which can be studied locally; the history of a local industry or craft, social relationships in a particular community, culture and dialect, change in the family, the impact of wars and strikes, and so on. An oral history project will be certainly feasible. It will also demonstrate very well, especially if the project focuses on the historical roots of some contemporary concern, the relevance of historical study to the immediate environment.

In schools, projects on children's own family history have been developed which provide an effective way of linking their own environment with a wider past. Family history has two other special educational merits. It assists a child-centred approach, for it uses as the project's basis the child's own knowledge of its family and kin and access to photographs, old letters and documents, newspaper cuttings, and memories. Equally, family history encourages the involvement of parents in school activity.

A child's own family history represents perhaps the simplest type of project subject. It is more suited to suggesting than to solving a historical problem. Older groups are likely to choose some issue of more collective interest. At Corpus Christi College, Oxford, for example, Brian Harrison led a group of his students in a small research study on the history of college servants, a group of workers whose old-fashioned deferential respect for their employers, loyalty, meticulousness in their craft, and formality of dress and manner, are understandably perplexing to most students today. Through the project they came to a better understanding of the college servants—and vice versa—and at the same time of the significance of history itself. As one commented: 'I found equally important and interesting . . . seeing the impact of social change in really close detail . . . how changes in the general social environment changed the style of life, values, and relationships within a traditional community.'[1] The immediate environment also gains, through the sense of discovery in interviews, a vivid historical dimension: an awareness of the past which is not just known, but personally felt. This is especially true for a newcomer to a community or district. It is one thing to know that streets or fields around a home had a past before one's own arrival; quite different to have received from the remembered past, still alive in the minds of the older people of the place, personal intimacies of love across those particular fields, neighbours and homes in that particular street, work in that particular shop.

Such fragmentary facts are not merely evocative in themselves, but can be used as the raw material for worthwhile history. It is possible for even a single student in a summer vacation project, with interviews, to make a useful extension of historical knowledge—and also to create new resources which others may be able to use later. With a group project the opportunities naturally enlarge. The number of interviews can be greater, the archival searches more extensive, the subject more ambitious.

The group project has some special characteristics of its own. Instead of the atmosphere of competition common in education, it requires a spirit of intellectual co-operation. Isolated reading, examinations, and lecture sessions give way to collaborative historical research. The joint inquiry will also bring teachers and students into a much closer, less hierarchical relationship, giving far more chance of informal contact between them. Their dependence will become mutual. The teacher may bring special experience in interpretation and in knowledge of existing sources, but will rely on the support of the students as organizers and field-workers. In these ways some of the students are likely to show unexpected skills. The best essay-writer is not necessarily the best interviewer—nor is the teacher. A much more equal situation is created. But, paradoxically, at the same time, by resolving—or at least suspending—the conflict between research and teaching, it enables the teacher to be a better professional. The group project is both research and teaching, inextricably mixed, and as a result each is done more effectively.

The essential value of both group and single projects is, however, similar. Students can share in the excitements and satisfactions of creative historical research of intrinsic worth. At the same time they gain personal experience of the difficulties of such work. They formulate an interpretation or theory and then find exceptional facts which are difficult to explain away. They find that the people whom they interview do not fit easily into the social types presented by the preliminary reading. They need facts, or people, or records which prove tantalizingly elusive. They encounter the problems of bias, contradiction, and interpretation in evidence. Above all, they are brought back from the grand patterns of written history to the awkwardly individual human lives which are its basis.

Both kinds of project also have the important consequence of taking education out of its institutional retreats into the world. Both sides gain from this. Interviewing can bring together people from different social classes and age groups who would otherwise rarely meet, let alone get to know each other closely. Much of the widespread hostility to students is based on little knowledge of what they are actually like or do, and these meetings can bring an appreciation of the serious-mindedness and idealism which is widespread among them. They can also show ordinary people that history need not be irrelevant to their own lives. Conversely, teachers and students can become more directly aware of the image which they present to the wider public. And through entering into the lives of their informants, they gain more understanding of values which they do not share, and often respect for the courage shown in lives much less privileged than their own.

Yet the nature of the interview implies a breaking of the boundary between the educational institution and the world, between the professional and the ordinary public, more fundamental than this. For the historian comes to the interview to learn: to sit at the feet of others who, because they come from a different social class, or are less educated, or older, know more about something. The reconstruction of history itself becomes a much more widely collaborative process, in which non-professionals must play a critical part. By giving a central place in its writing and presentation to people of all kinds, history gains immensely. And old people especially benefit too. An oral history project can not only bring them new social contacts and sometimes lead to lasting friendships; it can render them an inestimable service. Too often ignored, and economically emasculated, they can be given a dignity, a sense of purpose, in going back over their lives and handing on valuable information to a younger generation.

These changes made possible through oral history are not confined to the writing of books or projects. They also affect the presentation of history in museums, record offices, and libraries. These all now have a means of infusing life into their collections, and through this, of bringing themselves into a more active relationship with their community. They can set up their own research projects, like Birmingham's on the city's baths and washhouses and Southampton's on its West Indian port community, or the Imperial War Museum programmes on early aviation and on conscientious objectors. In recent years many British museums have been among sponsors of oral history projects giving short-term work to young unemployed through the Manpower Services Commission, in a way which recalls the Federal Writers Projects of the New Deal era in the United States. A particularly remarkable approach is the 'ecomusée' programme begun in Belgium by Étienne Bernard, with the ideal of a museum without a building, working through communities with recording projects and temporary exhibitions of photographs and objects from the neighbourhood, which are then returned to their owners. Interestingly, the Jewish Museum at Manchester was an outcome of the Manchester Studies oral history programme launched by Bill Williams from the Polytechnic, which stimulated the city's Jewish community to save a closed Victorian synagogue. It has been opened as a permanent museum where you can lift a phone as you look at a display of the objects given and listen to memories relating to them. At Erddig, a recently opened National Trust house in Cheshire, you enter through the servants' quarters to the sound of the voices of the last generation of servants and their masters.

Oral history research can also help bring the display itself closer to the historical original. The 'period setting' for objects is replaced by the reconstruction of a real room, with, for example, tools and shavings and half-made baskets left about as if the craftsman was still using it. Indeed, in some museums it will still be used from time to time; and there are a few, like the working farm museum at Acton Scott in Shropshire, where recording work and daily use of the old processes are the linked objective of the whole enterprise. When older local people see this kind of museum, they are likely to have comments, and may even help with improvements by offering articles of their own. In one lively East London museum, attendants who heard this kind of conversation would alert a curator, and the old people would immediately be offered a cup of tea and a chance to record some of their impressions on the spot. As in most community projects, some of the recordings will be used to make educational tapes for use in local schools; and weekends have been arranged for the school-children—normally Sixth Formers—to meet the old people. Thus an active dialogue develops between old people, their own local history, and a museum which has become a social centre. Here is a model of a social role for history with great potential, which needs to be taken up elsewhere.

The use of interviewing for historical presentation in broadcasting is of course long-standing. Here indeed is a fine tradition of oral history techniques which goes back many years—in fact well before the term 'oral history' was introduced. Professional historians are of course given their own chance for brief lectures in the intervals between programmes on Radio Three. But most of those I know show much more interest in those radio and television programmes which re-evoke history through the use of raw material, some of it dating from the original period, some recorded retrospectively. For the historian of the future the preservation of many of these programmes, along with others in the BBC Sound Archives, will provide a rich source. It is very unfortunate that at present, by contrast, only a very small proportion of what is being broadcast on television is being preserved, and historians have shown curiously little interest in this systematic destruction of records.

In historical broadcasting it is the introduction of people, the original actors, which brings the programmes alive. Some local radio stations have deliberately used this type of programme to encourage links and exchanges with their local community, through programmes of voices stimulating listeners to send in their own comments and offer to be interviewed in turn. A weekly series of this kind by Radio Stoke-on-Trent lasted for two years. The 'Making of Modern London' series on London Weekend Television has been linked with a competition for projects by viewers themselves, for which local schools and old people's centres as well as individuals entered. For their 'Making Cars' series, Television History Workshop opened a shop near the factory to gather materials, and also reactions to the programmes as broadcast. But perhaps the most impressive broadcasting experiment has been in Sweden. Here Bengt Jansson was able to organize through Swedish educational television a series on social change ('Bygd i förvardling'), concentrating on two regions of the country, where 700 local discussion circles were set up in association with the programmes, bringing together in all 80,000 people to join in exchanging their own experience of history in a lifetime.

There have also been attempts to use oral history in film making. In television and film a recurrent problem is that a series of interwoven interviews easily becomes visually repetitive, and—despite vivid moments—lacks dramatic action. When available, old film can provide one effective contrast. An alternative approach was used for the filming of Ronald Blythe's 'Akenfield', with local people re-enacting some of the scenes, their words unscripted. They gave their services to the film freely at weekends, and brought to each session clothes, props, and food. They would simply be warned in advance what the scene was to be about, so that they could meet in the appropriate dress and frame of mind. The result is certainly a remarkable, if somewhat puzzling film. It has moments which are deeply moving just because they are so ordinary: like the funeral sequence, with its awkward silences, the inadequacy of words when they come, the too slowly sung hymn, and afterwards the bad jokes and stories told again and again. But some who saw the film found it simply boring, ordinary, and without any obvious point. Above all it lacked a strong story.

This is one reason why some of the parallel experiments in drama have proved more successful. Angela Hewins's 'The Dillen' movingly records the life of George Hewins, a man who wrote with difficulty yet had a rare gift for words in telling his story, brought up as an orphan by his grandmother in a Stratford-on-Avon common lodging house, struggling for a living as a casual labourer, cruelly maimed in the trenches in the First World War. It was produced as a play by the Royal Shakespeare company, with a core of professional actors supported by a hundred and fifty local volunteers, who made their own costumes, and included a band; and the actual performance left the theatre itself to move round the town, stopping for scenes in a park, a building site, by the river, and on a disused rail track. Each night a crowd who soon outnumbered the original audience would gather and follow, and in the fairground atmosphere of the interval out in the meadows you could listen to groups of them exchanging their own memories of just these same places. After the shattering First World War scenes, huddled into a military tent, the performance would culminate in a torchlit peace procession of actors, audience, and bystanders, by now seven hundred strong, back into the town. This was an entirely new, and powerfully moving, form of community theatre. Several other companies, such as Age Exchange, have in a more modest way developed work based on oral history material for performances in village halls and community centres. In most work of this kind, however, although the words and even the acting have come from local people, the essential direction has remained in other hands. If there was a common purpose, it was one imposed from without.

An alternative approach to reminiscence theatre can be found in participatory drama like that developed by Elyse Dodgson in work for London schools, now based on the Royal Court Young People's Theatre. In this children have gathered material from their own families and then worked together in producing a joint performance—of impressively high quality. There are no doubt special difficulties in theatre and film because of the technical requirements and costs of the medium and its domination by an international professionalism. A similar problem applies in national broadcasting. Nevertheless, a choice, often difficult, has to be made in most other types of project, especially in education. For the co-operative nature of the oral history approach has led to a radical questioning of the fundamental relationship between history and the community. Historical information need not be taken away from the community for interpretation and presentation by the professional historian. Through oral history the community can, and should, be given the confidence to write its own history.

Some of the most interesting ventures in this direction have again come from Sweden, particularly through the role of the Swedish state exhibition-organization in encouraging local self-help exhibitions, and Sven Linqvist's book *Grav där du står* (1978) (*Dig Where You Stand*) which has provided a practical manual for workers to write the histories of their own workplaces—from their own standpoint, rather than that of employers and shareholders—combining both documentary and oral sources.

The full possibilities of the approach have, however, been still more strikingly revealed in Poland. It its true that the tape recorder has as yet made little impact there, so that the life-history movement, which dates back to a 'humanistic' tradition in Polish sociology established between the wars, still works through the encouragement of written autobiographical memoirs rather than oral testimony. No doubt this limits who can participate in it. Nevertheless, since 1945 memoir-writing has become an important form of popular self-expression in Poland, allowing discussion not only of pre-war society and the experience of Nazi occupation, but also of the radical social reconstruction which has taken place under subsequent Communist rule. The key to this success has been the use of memoir competitions, which are organized by the national newspapers and radio, and by local newspapers in every big city. Broad themes are set, and quite substantial prizes offered, two or three times a year. Each competition normally attracts a thousand or more entries. The best results are serialized in newspapers, and published as collections in book form. By now several hundred thousand Poles have entered for competitions, and a special national archive has been developed for the material collected. Popular memoir-writing, in short, has become a recognized part of the new national way of life, to an extent which has few parallels either in other Communist countries or in the West. And this Polish success in generating a kind of democratic enthusiasm for history has also led on to the forming of collective memoir-writing groups at some of the big factories, mines and steelworks. A sociologist may launch the initial meeting, and help with suggesting themes and later with the publication of books produced by the group, but the essential dynamic is provided by the commitment of the group members. Where else today could you find co-operative groups of industrial workers, up to 200 in number, helping to correct and enlarge their own life-history drafts through coming together regularly, after work, for two-hour discussion meetings?

A similar hope has inspired some of the relatively small British co-operative local oral history groups which have issued cheap cyclostyled broadsheets of transcribed extracts from recordings, adult education local history projects, and joint projects between oral historians and trade unionists. The springing up of such groups in every major city has indeed been one of the most striking features of the oral history movement: in London alone, the 1985 Exploring Living Memory exhibition held in the Festival Hall attracted 100,000 visitors in a fortnight, infusing it with a constant hubbub of talking as they saw and discussed the work of ninety projects from local history and publishing groups, hospitals and old people's centres, schools and so on. Among them, the most radical model has been provided by the People's Autobiography of Hackney. This arose from a group, originally connected with the WEA (Workers' Educational Association), which met in a local book and community centre called Centreprise. Members of the group varied in age from their teens to their seventies, but all lived in or near Hackney in East London. Their occupations were very mixed. The group was an open one, brought together by notices in the local papers, libraries, and other places. Any member could record anyone else. At the group meetings they played and discussed their tapes—sometimes also recording these discussions—and planned ways of sharing what was collected with a Hackney audience. For this reason they especially emphasized publishing and issued a series of cheap pamphlets, assisted by a local library subsidy, based on transcriptions and written accounts of people's lives, which have had a large local circulation. These pamphlets have in turn stimulated reactions from other people and led to more discussion and recordings. The group also collected photographs, and combined this material as tapes and slides for historical presentations to audiences in the community such as hospital patients and pensioners' associations—another way of giving back to people their own history, showing them it was valued, and stimulating their own contributions. The People's Autobiography thus aimed, on the one hand, to build up through a series of individual accounts a composite history of life and work in Hackney, and, on the other, to give people confidence in their own memories and interpretations of the past, their ability to contribute to the writing of history—confidence, too, in their own words: in short, in themselves.

This is not only true of old people who are still active and interested in life. Another very striking recent development of oral history has been reminiscence therapy. It has been increasingly recognized by specialists on ageing that reminiscing may be one important way in which old people keep their sense of self in a changing world. More remarkably, it can be used to rekindle the spirit of the acutely withdrawn and depressed, and even as a form of treatment for psychotic and dementing old people. The 'Recall' reminiscence tape and slide kit created by the oral historian Joanna Bornat at Help the Aged has sparked a growing movement, which we discuss in detail in Chapter 5, among professionals caring for the old.

The possibility of using history for such a constructive social and personal purpose comes from the intrinsic nature of the oral approach. It is about individual lives—and any life is of interest. And it depends upon speech, not upon the much more demanding and restricted skill of writing. Moreover, the tape recorder not only allows history to be taken down in spoken words but also to be presented through them. In a 'Recall' tape and slide show, or a museum demonstration of craft techniques, or a historical talk, the use of a human voice, fresh, personal, particular, always brings the past into the present with extraordinary immediacy. The words may be idiosyncratically phrased, but all the more expressive for that. They breathe life into history.

Something more is to be learned from them than mere content. Recordings demonstrate the rich ability of people of all walks of life to express themselves. George Ewart Evans has shown in his many books how the dialect of the East Anglian farm labourer, long scorned by the county landowning class for his notable inarticulacy, carries a Chaucerian grammatical and expressive strength which is hard to equal in conventional English. And this kind of discovery has been shared by oral historians wherever they have worked. The tape recorder has allowed the speech of ordinary people—their narrative skill for example—to be seriously understood for the first time. Educationists a few years ago, under the influence of Basil Bernstein, were assuming that working-class speech was a fatal handicap, a constraint which imprisoned all but the simplest types of thought. Now, with the help of tape recorders, the magazine *Language and Class Workshop* can challenge Bernstein's theories with its published transcripts; and in America 'urban folklore' has become an accepted literary genre. However, it may well be a long time before such revaluations reach general acceptance. Meanwhile, one of the key social contributions which can be made by the oral historian, whether in projects or through bringing direct quotation into written history, is to help give ordinary people confidence in their own speech.

In discovering such a purpose, oral historians have travelled a long way from their original aim—and there is, undoubtedly, some danger of conflict between the two. On the level of the interview itself, for example, there have been telling criticisms of a relationship with informants in which a middle-class professional determines who is to be interviewed and what is to be discussed and then disappears with a tape of somebody's life which they never hear about again—and if they did, might be indignant at the unintended meanings imposed on their words. There are clear social advantages in the contrasting ideal of a self-selected group, or an open public meeting, which focuses on equal discussion and encourages local publication of its results; and of individual recording sessions which are conversations rather than directed interviews. But there are also drawbacks in the alternative.

The self-selected group will rarely be fully representative of a community. It is much more likely to be composed from its central groups—people from a skilled working-class or lower middle-class background. The local upper class will rarely be there, nor will the very poor, the less confident especially among women, or the immigrant from its racial minority. A truer and socially more valuable form of local oral history will be created when these other groups are drawn in. Its publications will be much more telling if they can juxtapose, for example, the mistress with the domestic servant, or a millowner with the millworkers. It will then reveal the variety of social experience in the community, the groups which had the better or the worse of it—and perhaps lead to a consideration of what might be done about it. Local history drawn from a more restricted social stratum tends to be more complacent, a re-enactment of community myth. This certainly needs to be recorded, and a self-sufficient local group which can do this is undoubtedly helping many others besides itself. But for the radical historian it is hardly sufficient. History should not merely comfort; it should provide a challenge, and understanding which helps towards change. For this the myth needs to become dynamic. It has to encompass the complexities of conflict. And for the historian who wishes to work and write as a socialist, the task must be not simply to celebrate the working class as it is, but to raise its consciousness. There is no point in replacing a conservative myth of upper-class wisdom with a lower-class one. A history is required which leads to action: not to confirm, but to change the world.

In principle there is no reason why local projects should not have such an object, while at the same time continuing to encourage self-confidence and the writing of history from within the community. Most groups will normally contain some members with more historical experience. They certainly need to use tact; to undervalue rather than emphasize their advantage. But it is everybody's loss in the long run if they disown it: their contribution should be to help the group towards a wider perspective. Similar observations apply in the recording session where the essential need is mutual respect. A superior, dominating attitude does not make for a good interview anyway. The oral historian has to be a good listener, the informant an active helper. As George Ewart Evans puts it—'although the old survivors were walking books, I could not just leaf them over. They were persons.'[2] And so are historians. They have come for a purpose, to get information, and if ultimately ashamed of this they should not have come at all. A historian who just engages in haphazard reminiscence will collect interesting pieces of information, but will throw away the chance of winning the critical evidence for the structure of historical argument and interpretation.

The relationship between history and the community should not be one-sided in either direction: but rather a series of exchanges, a dialectic, between information and interpretation, between educationists and their localities, between classes and generations. There will be room for many kinds of oral history and it will have many different social consequences. But at bottom they are all related.

Oral history is a history built around people. It thrusts life into history itself and it widens its scope. It allows heroes not just from the leaders, but from the unknown majority of the people. It encourages teachers and students to become fellow-workers. It brings history into, and out of, the community. It helps the less privileged, and especially the old, towards dignity and self-confidence. It makes for contact—and thence understanding—between social classes, and between generations. And to individual historians and others, with shared meanings, it can give a sense of belonging to a place or in time. In short, it makes for fuller human beings. Equally, oral history offers a challenge to the accepted myths of history, to the authoritarian judgement inherent in its tradition. It provides a means for a radical transformation of the social meaning of history.

# Historians and Oral History

THE extensive modern use of the term 'oral history' is new, like the tape recorder; and it has radical implications for the future. But this does not mean that it has no past. In fact, oral history is as old as history itself. It was the *first* kind of history. And it is only quite recently that skill in handling oral evidence has ceased to be one of the marks of the great historian. When the leading professional historian of mid-nineteenth-century France, Jules Michelet, professor of the École Normale, the Sorbonne, and the Collège de France, and chief historical curator of the National Archives, came to write his *History of the French Revolution* (1847–53), he assumed that written documents should be but one source among many. He could draw on his own memory: he had been born in Paris in 1798, within a decade of the fall of the Bastille. But for ten years he had also been systematically collecting oral evidence outside Paris. His intention was to counterbalance the evidence of official documents with the political judgement of popular oral tradition:

When I say *oral* tradition, I mean *national* tradition, which remained generally scattered in the mouths of the people, which everybody said and repeated, peasants, townsfolk, old men, women, even children; which you can hear if you enter of an evening into a village tavern; which you may gather if, finding on the road a passer-by at rest, you begin to converse with him about the rain, the season, then the high price of victuals, then the times of the Emperor, then the times of the Revolution.

Michelet was clearly skilled at listening, and drawing an informant out. He also had distinct ideas about the areas in which oral evidence was more, or less, reliable. As a scholar in his own time he was exceptional; but he was certainly not peculiar. Yet within a century the historical profession had so far turned its back on its own traditional skills, that Professor James Westfall Thompson commented on Michelet's passage, in his monumental *History of*

*Historical Writing* (1942), 'this may seem like a strange way of collecting historical data'.[1] How did this reversal come about? What were the stages by which oral history lost its original eminence?

One of the underlying reasons becomes clear as soon as we look at the scope of oral tradition in pre-literate societies. At this stage all history was oral history. But everything else had to be remembered too: crafts and skills, the time and season, the sky, territory, law, speeches, transactions, bargains. And oral tradition itself was very varied. Jan Vansina in his classic *Oral Tradition: a study in historical methodology* (1965),[2] divided African oral tradition into five categories. First there are formulas—learning formulas, rituals, slogans and titles. Next there are lists of place names and personal names. Then come official and private poetry—historical, religious, or personal. Fourthly there are stories—historical, didactic, artistic, or personal. Lastly there are legal and other commentaries. Not all of these can be found in all African societies. Official poetry and historical stories, for example, arise only with a relatively high degree of political organization. Nevertheless, in most societies there is normally a considerable range of oral evidence. The social importance of some of these oral traditions also resulted in reliable systems for handing them down from generation to generation with a minimum of distortion. Practices such as group testimony on ritual occasions, disputations, schools for teaching traditional lore, and recitations on taking office, could preserve exact tests through the centuries, including archaisms even after they had ceased to be understood. Traditions of this type resemble legal documents, or sacred books, and their bearers become in many African courts highly specialized officials. In Rwanda, for example, genealogists, memorialists, rhapsodists, and *abiiru* were each responsible for the preservation of different types of tradition. The genealogists, *abacurabwenge*, had to remember the lists of kings and queen mothers; the memoralists, *abateekerezi*, the most important events of the various reigns; the rhapsodists, *abasizi*, preserved the panegyrics on the kings; and the *abiiru* the secrets of the dynasty. 'Without us the names of kings would vanish into oblivion, we are the memory of mankind', the praise

singers justly claimed: 'I teach kings the history of their ancestors
so that the lives of the ancients might serve them as an example,
for the world is old, but the future springs from the past.'[3]

There were also village tradition-bearers, who, more often
than the court specialists, have continued to hand down traditions into the present. They had their equivalents in many other
cultures, as in the Scandinavian 'skald' or the Indian 'rajput'. A
dramatic encounter with one such West African 'griot' has been
described by Alex Haley in his account of the rediscovery of his
own ancestry—subsequently given great publicity in the semi-
fictionalized form of *Roots* (1976). His family had a tradition
rare among Black Americans—of how their first ancestor came
to the colonies as a slave, including a few details: how he had been
captured when chopping wood, his African name had been Kin-
tay, he called a guitar a 'ko' and a river 'Kamby Bolongo'; how he
had landed at 'Naplis' and worked with the English name Toby
for Mas' William Waller. For this Black family descent in
America itself, Haley was able to provide proof from archival
researches, down to an advertisement in the *Maryland Gazette* of
October 1767 for 'fresh slaves for sale' of the *Lord Ligonier* and a
transfer deed between the brothers John and William Waller of
'one Negro man slave called Toby'. But all this followed the high
moment of his search, back across the Atlantic—a moment in
which it now seems enthusiasm may have gone further than the
evidence warranted. His ancestor's language had been identified
as Mandinka and 'Kamby Bolongo' as the Gambia River; and
then, in Gambia, he found that there was an old family clan
called Kinte. So far so good. Then after a search, a tradition-
bearer of the clan, or 'griot', was located in a tiny, distant hamlet
in the interior. Accompanied by interpreters and musicians, Alex
Haley eventually reached him: 'And from a distance I could see
this small man with a pillbox hat and an off-white robe, and even
from a distance there was an air of "somebodiness" about him.'
The people gathered around Alex Haley in a horseshoe to stare at
the first black American they had seen. And then they turned to
the old man:

The old man, the *griot*, the oral historian, Kebba Kanga Fofana, 73 rains
of age, began now to tell me the ancestral history of the Kinte clan as it

had been told down across the centuries, from the times of the forefathers. It was as if a scroll was being read. It wasn't just talk as we talk. It was a very formal occasion. The people became mouse quiet, rigid. The old man sat in a chair and when he would speak he would come up forward, his body would grow rigid, the cords in his neck stood out and he spoke words as though they were physical objects coming out of his mouth. He'd speak a sentence or so, he would go limp, relax, and the translation would come. Out of this man's head came spilling lineage details incredible to behold. Two, three centuries back. Who married whom, who had what children, what children married whom and their children, and so forth, just unbelievable. I was struck not only by the profusion of details, but also by the biblical pattern of the way they expressed it. It would be something like: 'and so and so took as a wife so and so and begat and begat and begat', and he'd name their mates and their children, and so forth. When they would date things it was not with calendar dates, but they would date things with physical events, such as . . . a flood.

So step by step the old man recounted the history of the Kinte clan: how they had come out of Old Mali, had been blacksmiths, potters, and weavers, had settled in the present village, until, roughly between 1750 and 1760, a younger son of the family, Omoro Kinte, took a wife, Binta Kebba, by whom he had four sons, whose names were Kunta, Lamin, Suwadu, and Madi.

By the time he got down to that level of the family, the *griot* had talked for probably five hours. He had stopped maybe fifty times in the course of that narrative . . . And then a translation came as all the others had come, calmly, and it began, 'About the time the king's soldiers came'. That was one of those time-fixing references. Later in England, in British Parliamentary records, I went feverishly searching to find out what he was talking about, because I had to have the calendar date. But now in back country Africa, the *griot* Kebba Kanga Fofana, the oral historian, was telling the story as it had come down for centuries from the time of the forefathers of the Kinte clan. 'About the time the king's soldiers came, the eldest of these four sons, Kunta, went away from this village to chop wood and was seen never again.' And he went on with his story. I sat there as if I was carved of rock . . .

Alex Haley did, after a few moments, pull out his own notebook, and show the interpreters that this was the same story that he had himself heard as a child from his grandmother on the front porch of her house in Tennessee; and there then followed a spontaneous ceremony of reconciliation with his own people, in which he laid hands on their infants, and they took him into their mosque and prayed in Arabic, 'Praise be to Allah for one long lost from us whom Allah has returned.'[4]

For a number of reasons, the identification of Kinte is much more doubtful than Alex Haley believed in that moment. His 'griot', who lacked the full traditional training, was not an ideal tradition-bearer, but like a good 'griot' was searching the genealogical store in his mind for the evidence needed for an audience—and he may have had an idea in advance of what Haley wanted. Subsequently, there have been variations in minor detail when he has repeated his testimony. More important, the African and American generations fit awkwardly—although this could be due to a telescoping not uncommon in oral tradition— and the time-fixing reference is very weak for an area in which European soldiers had been present for a long time. But we can easily find other instances of accuracy of oral tradition in non-literate societies if we look elsewhere; for example, to ancient Greece, where the accuracy of description of details of obsolete armour and name-lists of abandoned cities, preserved orally for six hundred years before the first written versions of the *Iliad* were circulated, has been vindicated by classical scholarship and archaeology.

Nevertheless, Haley's story does bring home with rare power the *standing* of the oral historian before the spread of documentation in literate societies made redundant such public moments of historical revelation. We can no longer distinguish, like the Swahili, between the 'living dead', whose names are still recalled in oral tradition, and the absolutely forgotten. The modern genealogist works in private silence in a record office. Memory is demoted from the status of public authority to that of a private aid. People still remember rituals, names, songs, stories, skills; but it is now the document which stands as the final authority, and the guarantee of transmission to the future. Consequently, it is just those public and long-term oral traditions, which were once the most prestigious, which have proved most vulnerable. By contrast, personal reminiscence and private family traditions, which are rarely committed to paper just because most people do not think them of much importance to others have become the standard type of oral evidence. And it is normally only among social groups of low prestige, such as children, the urban poor, or isolated country people, that other oral traditions such as games, songs, ballads, and historical stories are now collected. And the strongest communal memories are those of beleagured outgroups. The Gaelic-speaking crofting communities of north-west Britain remember the eighteenth-century Highland sheep clearances which drove them of their old townships to the sea's edge, as if they had been yesterday. In France the royalist families of the Vendée handed down their story of resistance to the republic for 150 years. Still more remarkably, in the Protestant mountain valleys of the Cévennes today, family traditions still yield a more accurate interpretation than contemporary documents of the unprecedented—and hence misreported—guerrilla war of the Camisards ('whiteshirts') in 1702–4, in which their peasant ancestors successfully held at bay the royal army of Louis XIV and secured the survival of their faith. The changing social standing of the bearers of oral tradition is thus clearly related to its long-term decline in prestige—and, conversely, to its current radicality.

In Western Europe this came about very slowly. The first written histories probably go back 3,000 years. They set down existing oral tradition about the distant past and gradually also began to chronicle the present. Just because it began so early in Europe, this stage is more easily observed where it happened more recently: in the systematic collecting of historical traditions from commoners by the third-century Chinese royal historian Sima Qian and from noble families ordered by the Japanese emperor in the eighth century, the assembling of memories of the prophets in the ninth-century Muslim world, or the precious documentation of pre-conquest Aztec history and culture from the memory of old men by Sahagun and the Spanish Franciscan friars in mid-sixteenth-century Mexico. But we know that from quite an early stage there were a few outstanding European historical writers who tried to evaluate their evidence. The method of Herodotus, for example, in the fifth century BC was to seek out eyewitnesses and cross-question them. By the third century

AD we can find Lucian advising the would-be historian to look
for his informant's motives; while Herodian cites enough of his
sources to suggest the order in which he rates them—antiquarian
authorities, palace information, letters, senate proceedings, and
other witnesses. And in the early eighth century Bede, in the
preface to his *History of the English Church and People*, carefully distinguished his sources. For most of the English provinces
he had to rely on oral traditions sent to him by other clergy, but he
was able to draw on the records at Canterbury, and he even
secured copies of letters from the papal archives through a
London priest who visited Rome. But he was surest of the evidence
for his own Northumbria, where 'I am not dependent on any one
author, but on countless faithful *witnesses* who either know or
remember the facts, apart from what I know myself'.[5]

Bede's attitude to evidence, and his assumption that he could
be most trusted where he had been able to collect oral evidence
from eyewitnesses himself, would have been shared by all the
most critical historians into the eighteenth century—not to
mention the many less meticulous chroniclers and hagiographers
who stood between them. Neither the spread of printing, nor the
secular rationality of the Renaissance, brought any changes in this
way. This is perhaps less surprising when it is realized that the
typical scholar *heard*, rather than himself *read*, the printed books
which became available. And when the truth mattered most, it
had to be spoken. The popes pronounced their final words on
Catholic doctrine *ex cathedra*: and in both the Christian and
Muslim worlds the courts—which had quickly enough discovered
how easy it was to forge a written charter—continued to insist
that witnesses must be heard, because only then could they be
cross-challenged.[6] Even accounts had to be checked aloud, or
'audited', each year. And in practice the best-known historians
remained rather less careful than Bede. Guiccardini in sixteenth-
century Italy, for example, avoids the direct quotation of documents, and assumes his own participation in the times he describes
is a sufficient guarantee of truth. Clarendon's *History of the
Rebellion and Civil Wars in England* (1704) carries a similar tone,
although he does occasionally refer to reminiscence, and he did
trouble to look at the journals of the House of Commons for the

ten years when he was not a Member. Bishop Burnet's *History of His Own Time* (1724) is less magisterial, but again assumes the prime value of oral evidence, which he handles with a notable care. He cites the authors of his stories regularly, and when his witnesses disagree he sets them against each other. Printed authorities, by contrast, he assumes to be inferior: 'I leave all common transactions to ordinary books. If at any time I say things that occur in any books, it is partly to keep the thread of the narration in an unentangled method.'[7]

It is perhaps more surprising to find little immediate change, at least in the attitude to evidence for recent history, amongst the historians of the eighteenth-century Enlightenment. Voltaire was certainly cynical enough about the 'absurd' myths of oral tradition from the remote past, recited from generation to generation, which had been the original 'foundations of history': indeed, the remoter their origin, the less their value, for 'they lose a degree of probability at every successive transmission'. He rejoiced that 'omens, prodigies, and apparitions are now being sent back to the regions of fable. History stood in need of being enlightened by philosophy.' From modern historians he demanded 'more details, better ascertained facts'. But although for his own works he collected both oral and documentary evidence, he rarely cited his sources and his general comments suggest a lack of distinction between them. He boasted in his *History of Charles XII* (1731), for example, that he had 'not ventured to advance a single fact, without consulting eyewitnesses of undoubted veracity'. After its publication, he cited as an indication of his reliability a letter of approval from the king of Poland, who 'himself had been an eyewitness' of some of the events described. He also defended his failure to cite authorities in *The Age of Louis XIV* (1751) on the ground that 'the events of the first years, being known to every one, wanted only to be placed in their proper light; and as to those of a later date, the author speaks of them as eyewitness'. By contrast he did feel a need in his *History of the Russian Empire under Peter the Great* (1759–63) to name, at least at the start, 'his vouchers, the principal of which is Peter the Great himself'.[8] For this work he had the assistance of documents selected and copied by the Russian officials and sent to his home in Geneva. Voltaire, while retaining a special regard for personal witness, reveals curiously little awareness of the possible bias either in a monarch's own judgement of his reign, or in a set of documents preserved and even selected by the royal officials themselves.

Voltaire was, moreover, a historian with many distinguished admirers. James Boswell recorded a breakfast discussion in 1773 between Samuel Johnson, who had left the codifying of the English language and the delights of London to seek the direct experience of a primitive society in the Scottish islands, and two leaders of the Edinburgh Enlightenment, the lawyer Lord Elibank, and the philosopher-historian William Robertson, Principal of the university. The conversation turned to the last great revolt of the Scottish Highlands against English rule, the 1745 rebellion. Johnson agreed that this 'would make a fine piece of history', but countered Elibank's doubt 'whether any man of this age could give it impartially' by citing Voltaire's method in his *Louis XIV*: 'A man, by talking with those of different sides, who were actors in it, and putting down all that he hears, may in time collect the materials of a good narrative. You are to consider, all history was at first oral.' And he was firmly backed by the Scots historian, who also knew Voltaire: 'It was now full time to make such a collection as Dr Johnson suggested; for many of the people who were then in arms, were dropping off; and both Whigs and Jacobites were now come to talk with moderation.'[9]

It is no accident that this remarkable early call for an 'oral history' project came at this moment. They stood at the edge of a period of great change in the nature of historical scholarship. Behind it lay the cumulative effects of two centuries of printing: an explosion in historical resources which was both quantitative and qualitative. We may take, for an example, *A New Method of Studying History: recommending more easy and complete instructions for improvements in that science*, published by Langlet du Fresnoy, librarian to the Prince of Savoy, in 1713, and subsequently translated into Dutch, German, and English. As it happens there is nothing very new in the method itself which Fresnoy puts forward—he even asserts that those historians who combine 'hard study, and a great experience of affairs', are considerably superior to those 'that shut themselves up in their closets to examine there, upon the credit of others, the facts which themselves were not able to be informed of'.[10] Much more remarkable is his second volume, for it consists entirely of bibliography, listing altogether some 10,000 titles of historical works in the major European languages. The production of such a list indicates a substantial community of scholars. It also shows the development of basic professional resources. An English historian, for example, could now make use of a series of county and local histories, biographies and biographical collections, and travellers' accounts. Printed sets of church inscriptions, manuscript chronicles, and medieval public rolls were being published. In Bishop William Nicolson's *English Historical Library* he had available a critical bibliography. The apparatus for writing history from the closet was being assembled: it was becoming possible for some historians at least to dispense with their own field-work, and rely on documents and oral evidence published by others.

Nevertheless, the immediate effect of the immense expansion of printed sources which continued through the eighteenth century was a positive enrichment of historical writing. Voltaire could reasonably insist that a good modern historian pay 'more attention to customs, laws, mores, commerce, finance, agriculture, population. It is with history as it is with mathematics and physics. The scope has increased prodigiously.'[11] One can see the long-term impact of change particularly well in Macaulay, whose *History of England* (1848–55) was in terms of sales probably the most popular nineteenth-century history book in the English language. As a practising politician, and a master of style, Macaulay might be seen as an heir to Guiccardini and Clarendon. But perhaps the most brilliant passages of his book are those in which he gives the social background, from the way of life of the country squire to the condition of the urban and rural poor. He uses as his raw materials contemporary surveys, poetry and novels, diaries, and published reminiscences. He also makes particularly interesting use of oral tradition. In stories of the highwaymen who 'held an aristocratical position in the community of thieves', anecdotes 'of their ferocity and audacity, of their occasional acts of generosity and good nature, of their amours, of their miraculous escapes . . . there is doubtless a large mixture of fable;

but they are not on that account unworthy of being recorded; for it is both an authentic and an important fact that such tales, whether false or true, were heard by our ancestors with eagerness and faith'. He quotes at length a broadside street ballad which he calls 'the vehement and bitter cry of labour against capital', and argues that evidence of this kind must be used for social history. 'The common people of that age were not in the habit of meeting for public discussion, or haranguing, or of petitioning parliament. No newspaper pleaded their cause . . . A great part of their history is to be learned only from ballads.'[12]

As a general historian, Macaulay drew not simply on a wider range of published sources, but also on the development of a whole series of other modes of historical writing. One of the authorities he cited in using oral tradition was Sir Walter Scott. As a young man, before he began writing novels, Scott was a Border Country lawyer, and one of his first publications was a *Minstrelsy of the Scottish Border* (1802), a set of popular ballads which he had collected from country people with his friend Robert Shortreed. His own interest had in turn been partly awakened by a still earlier collection, Bishop Percy's *Reliques of Ancient English Poetry* (1765). But he could have chanced on others. Perhaps best known was William Camden's *Britannia* (1586), which includes chapters on the development of the English language, proverbs, and names as well as poetry. It is one of the founding works of the historical study of language and folklore. There was also the contrastingly radical work of the Newcastle populists, John Brand and Joseph Ritson, who saw the study of popular culture as a duty of 'the friends of man', and combined the collecting of oral tradition with schemes for encouraging popular self-expression in a simplified spelling of English based on the spoken vernacular language.[13]

Scott went on to make a still more important contribution to a second new form of historical writing, the historical novel. Here again he collected much of the oral evidence which he needed himself. He visited the Highlands, 'talking to Jacobites who had taken part in the '45 Rebellion'. Scott recognized through conversing with these old men what had really happened as a result of the '45. Culloden saw the end of a culture; the dispersal or the destruction of the Highland clans, a tribal society, and an older, fundamentally different way of life. 'The old men he talked to were truly historical documents; and contact with them helped to give his writing that veracity which informs earlier novels like *Waverley*, *The Antiquary*, *Rob Roy*, and *Guy Mannering*.' It was to honour his sources as much as to chaff himself that he prefaced some of his novels with Robert Burns' warning lines:

> A chiel's amang you takin' notes
> An' faith he'll prent it.[14]

Both as a note-taker, and in the form of the historical novel itself, Scott set the pattern for some of the major imaginative works on the nineteenth century. Dickens, for example, deliberately set many of his novels in the London world which he could remember from childhood, and when he could not draw easily on oral memory, as for *Hard Times*, set out for special field-work. Charlotte Brontë's *Shirley* draws much of its drama from her knowledge of local memories of the Luddite rising. George Moore's life-story of a domestic servant, *Esther Waters*, owes its realism to his habitual chatting below stairs in country houses and elsewhere. George Borrow came to understand the East Anglian gypsies in a similar way. In France, the work of Émile Zola sought the material for *Germinal* from his talks with the miners of Mons. Later on in Britain, Arnold Bennett was another great note-taker, and his *Clayhanger* was again a reconstruction of a remembered world. Closer still to Scott was Thomas Hardy, with his shrewd observation of traditional country customs, and ability to use them as illustrations of conflict and change within the whole social structure. But this is looking ahead—and to a stage when, to their own loss, historians were less prepared to learn from novelists.

A third type of historical work which had expanded especially fast from the end of the seventeenth century was the biographical memoir. In this the use of oral evidence remained, of course, an assumed method. The growing popularity of memoirs brought interesting extensions in scope. First, there were a number of projects for collections of biographies which aimed to represent whole social groups, rather than simply exceptional individuals.

The most famous of these projects, John Aubrey's *Brief Lives*, although known in his lifetime, was not in fact published until two centuries later, in 1898. Aubrey, who wrote that from boyhood 'he did ever love to converse with old men, as Living Histories', was an impoverished country gentleman, forced to turn his hobby into a living as an antiquarian research assistant working for others.[15] In the course of this he found time to put together stories and information from innumerable sources to compose a biographical portrait of his social circle, the seventeenth-century intelligentsia, as a whole. A more obscure example on a local level was Richard Gough's *Human Nature Displayed in the History of Myddle* (1833), in Shropshire, which had been written in 1700–6, and has recently attracted the interest of historians. In his preface to its republication W. G. Hoskins calls it 'a unique book. It gives us a picture of seventeenth-century England in all its wonderful and varied detail such as no other book that I know even remotely approaches.' Gough started by discussing the buildings of the parish; but once he reached the parish church he used its pews as the framework for a social survey, taking each pew-holding family in turn, discussing their origins and their occupations, and relating with relish either their successes, or their failings—drink, bribery, and whoring. This information, moreover, is not merely illustrative; for its value has also emerged, in a modern historical study, in establishing basic demographic facts, and correcting the misinterpretations which would otherwise have been made from more conventional sources such as wills and registers.[16] In the frankness with which he documented scandal, Gough is perhaps unique; but his focus on people rather than institutions provides one of the first instances of a valuable minority form of local history. A later example is the *History and Traditions of Darwen and its People* which J. G. Shaw, the editor of a local newspaper, recorded in shorthand and published from an old man in the town in 1889.

Still more striking, and undoubtedly a reflection of the early social and political emergence of the working class in Britain, was the remarkable nineteenth-century flowering of a very varied individual working-class autobiography: intellectual, political, or personal. It had several sources. One was the life published

as a moral example. The religious autobiographies of mid-
seventeenth-century Puritan sectarians were the first from the
lower classes; and the groups of *Spirituall Experiences* published
included, still more rarely, some testimonies by women. Stories
of conversion and rescue were again collected in the eighteenth
century from the Protestant Camisards in France and from old
dissenters and Methodist pioneers in Britain: and in the 1820s one
local historian of northern Wesleyanism not only secured a reso-
lution by Conference that it should be a duty of every superinten-
dent to collect testimonies of zeal and sufferings from early
Methodists, but chose as the frontispiece to his own book a
sketch (by himself) of 90-year-old Richard Bradley who had been
one of his own 'living oracles'.[17] Other mid-nineteenth-century
lives were edited by religious pamphleteers, introduced by
parsons, or given titles like *The Working Man's Way in the
World*. Morality was secularized by Samuel Smiles, who published
biographical collections of engineers, ironworkers, and tool-
makers as well as his classic, *Self-Help: with Illustrations of
Character and Conduct* (1859).[18] The notably full, early autobiography of the self-improving tailor Thomas Carter was
published by the moral and educational popularizer Charles
Knight in 1845 in much the same spirit. Quite a different vein was
represented by the memoir of picaresque adventure. In the eighteenth century this normally implied gambling or sexual intrigue,
but it could be extended into other forms of 'low life', and
circusmen's or poachers' autobiographies later carried some of
the same flavour.

There was a convergence of these two autobiographical
approaches in the mid-nineteenth century, as the working classes
made their political presence felt, and came to be seen as a
problem. The semi-autobiographical works of the journeyman
engineer, Thomas Wright—*Some Habits and Customs of the
Working Classes* (1867), *The Great Unwashed* (1868), and *Our
New Masters* (1873)—provided information for the middle class
which was comforting as well as colourful. There are signs too of
a concern in some authors to retain something of the liveliness of
working speech forms in print. And at the same time the working-
class movement itself began to produce autobiography, with the

early *Memoir of Thomas Hardy* (1832), on the French revolutionary years, followed before long by classics such as Samuel Bamford's *Early Days* (1848) and Chartist autobiographies like *The Life and Times of William Lovett* (1876); although the labour political biography eventually settled into a rather narrow form. The early emergence of working-class autobiography in Britain can therefore be linked closely to working-class activity, first in religion and then in politics. The same is true rather later on in France. It is striking that by contrast in Germany no tradition either of the social novel or of working-class autobiography was established in the nineteenth century. Only in 1904 did the socialist deputy in the Reichstag, Paul Göhre, launch the first series of autobiographies with the deliberate intention of revealing to middle-class readers both the conditions of ordinary life, and that working-class people shared 'human thoughts and feelings, and reacted to joy and suffering in the same way they did'.[19]

Lastly, among the new forms of historical writing can be seen, towards the end of the eighteenth century, the beginnings of an independent social history. At this stage there was no professional separation between the processes of creating information, constructing social theory and historical analysis, so that they proceed sometimes together, sometimes apart. One cannot, as a result, separate the origins of an 'oral history' method from general developments in the collection and use of oral evidence. Two of the earliest achievements, for example, came from Scotland. In 1781 John Millar published his *Origin of the Distinctions of Ranks*, which puts forward a historical and comparative theory of inequality. He not merely anticipated Marx by linking the stages in master–servant relationships with changes in economic organization, but produced in his discussion 'of the rank and condition of women in different ages' one of the first historical explanations of sexual inequality. This pioneering exercise in historical sociology depended on a wide variety of published sources from ancient histories to the recent descriptions of local social customs by European travellers in other continents. Then years later came a major step in the creation of source material, the first *Statistical Account of Scotland* (1791–9), a national collection of contemporary and historical social information carried out through the parish clergy and edited by Sir John Sinclair. There had been no investigation on a comparable scale in the British Isles since Domesday. Meanwhile, in England, one important model of social investigation was provided by the 'field-work' travels of Arthur Young, bringing together both his own observations and interviews with others in his influential reports on the state of British agriculture. William Cobbett's later travels, documenting the often devastating social consequences of economic progress in agriculture, used the same method in reply to Young. Others, less energetic, devised short-cuts which were to prove key methodological devices for the future. The first questionnaire has been attributed to David Davies, a Berkshire rector, who was investigating farm labourers' budgets, and sent out printed abstracts to potential collaborators, whom he hoped might collect similar information in other places. And it was for another investigation of *The State of the Poor*, again in the 1790s, that Sir Frederick Eden sent out one of the first modern interviewers: 'a remarkably faithful and intelligent person; who had spent more than a year in travelling from place to place, for the express purpose of obtaining exact information, agreeably to a set of queries with which I furnished him.'[20]

The nineteenth century was to see this process of development in field-work method, historical analysis, and social theory carried rapidly forward—but in a context of increasing separation and specialization. This was even true within field-work methodology itself. The travelling investigation, for example, became a field-work specialism of the colonial anthropologist, and the survey of the sociologist of 'modern' societies. And sharp differences emerged even between the form of survey method used in different European countries. In France, Belgium, and Germany, as well as in Britain, the survey was first used by independent philanthropists, medical reformers, and sometimes newspapers, and then taken up for official government investigations. But when the French began their first large-scale 'enquête ouvrière' under fear of the revolutionary uprisings of 1848, they did not seek evidence directly, but through their well-organized local bureaucracy. And the German social surveys which were begun in the 1870s were invariably sent out to local officials, clergy, teachers, or landowners, for return in essay form, following the model of the French and Belgian 'enquêtes'.

In Britain, by contrast, techniques for the direct collection of evidence were adopted. This began regularly with the launching in 1801 of the decennial census, carried out under central instructions by investigators dispersed throughout the country— thus establishing the national interview survey. Only the sum findings of the census were published. But the parliamentary social inquiries and Royal Commissions which increasingly came to be published as Blue books were also commonly conducted through interviewing, although of a different kind. Sometimes an on-the-spot investigation was made, but normally witnesses were summoned before the inquiring committee and questioned by them. The exchanges and arguments between the committee and witnesses were often reproduced along with the publication of the official report. They constitute a rich repository of autobiographical and other oral evidence. And their potential as source material was quickly realized. The Blue books were the basis of Disraeli's descriptions of working-class life in *Coningsby* and *Sybil*. And they proved equally useful to Karl Marx.

Marx and Engels, in their more immediate political writings, normally drew substantially both on direct experience of their own, and on reports, written and oral, from their innumerable correspondents and visitors. Equally, Engels' *Condition of the Working Class in England in 1844* combines material from newspapers, Blue books, and other contemporary comment with his own eyewitness accounts of working-class life. Engels had come to Manchester in 1842 to work in the English branch of his father's firm, and in his spare hours from the cotton mill was able to explore the industrial conditions of the city and to meet, with the help of a working-class girl, Mary Burns, some of the Chartist leaders. For his culminating theoretical analysis, however, Marx relied on published source material. *Capital* is heavily documented with both bibliography and footnotes. Apart from occasional quotations from classical literature, Marx cites two types of source; contemporary economic and political theory and comment; and contemporary description, often including vivid anecdotes, from newspapers and from the parliamentary Blue books. No doubt this decision of Marx to use only already published oral material, rather than carry out any new field-work, was partly due to personal taste, and partly to enable him to buttress his arguments with unassailable authorities. But given the influence which *Capital* was to have on the future of social history, it set a key precedent.

It is equally significant of the changing situation that such a choice was open to Marx. For we have still not exhausted the major new steps in the creation of oral source material for social history. In addition to the investigations of the government, social survey work was undertaken by voluntary bodies. By the late 1830s there were Statistical Societies in London, Manchester, and other cities, composed mainly of doctors, prosperous businessmen, and other professionals, which made important contributions to the techniques of collecting and analysing social information. They carried out local inquiries into working-class conditions, making pioneer use of the door-to-door questionnaire survey by paid interviewers, and publishing their findings in statistical tables prefaced by a brief report. In this form most of the original interview evidence was suppressed.

On the other hand, an alternative model was created by the newspaper investigation, which was developed in the 1840s, and culminated in the *Morning Chronicle* survey under Henry Mayhew. This inquiry, conceived in the wake of the great cholera epidemic of 1849, has been called 'the first empirical survey into poverty *as such*'.[21] Mayhew's aim was to demonstrate the relationship between industrial wage levels and social conditions. Instead of a door-to-door survey he therefore analysed a series of trades through a strategic sample. In each trade he looked for representative workers at each job level, and then took supplementary information from unusually well-paid workers at one extreme and distressed casual workers at the other. He obtained his information both from correspondence and by direct interview, and for both he gradually developed a detailed schedule of questions. Most striking was his actual interview technique. He seems to have felt a respect for his informants which was very rare among investigators of his time. His comments show both emotional sympathy and a willingness to listen to their views. Indeed, his changing standpoint shows that he was genuinely prepared to be influenced by them. No doubt this attitude helped him to be accepted into working-class family homes and receive their life-stories and feelings. And, significantly, it was linked to an unusual concern with their exact words. He normally went to interviews accompanied by a stenographer, so that everything said could be directly recorded in shorthand. And in his reports he gave very substantial space to direct quotation. In Mayhew's pages, as nowhere else, one can hear the ordinary people of mid-Victorian England speaking. It is because of this that they continue to be read.

Despite his popularity, Mayhew had no direct successors. But with the rise of the socialist movement in the late nineteenth century, a new concern to understand both the conditions and spirit of the working classes was felt both in Britain and in Germany. One result was the 'settlement' movement, which encouraged idealistic middle-class men and women to live among the poor, sometimes in groups as voluntary workers, but also alone, and even in disguise. In England, for example, a number of 'glimpses into the abyss' inside common lodging-houses and workhouses were written besides the famous accounts of Jack London and, later, George Orwell. In Germany in 1890 Paul Göhre, as a young theology student, worked incognito in a Chemnitz machine tool factory to produce *Three Months in a Workshop* (1895): a study of factory culture which marked a turning-point in German social inquiry—as well as setting Göhre on the path which later led him, as we have seen, to launch the first German working-class autobiographies. Robert Sherard also used clandestine techniques for his vivid accounts of industrial conditions in *The White Slaves of England* (1897): 'the factories I visited were visited by me as a trespasser, and at a trespasser's risk'. He generally avoided contact with employers, finding that they just laughed at his 'stories of grievances' in their 'luxurious smoking-rooms'. A similar direct understanding of working-class culture was openly sought by Alexander Paterson, whose *Across the Bridges* (1911) is based on his years living in South London. In rural studies it is expressed in the respect for country people of George Sturt's *Change in the Village* (1912), and still more in Stephen Reynolds' books on the Devon fisher-people with whom he shared a house, *A Poor Man's House* (1909) and *Seems So!* (1913). Reynolds' sympathy was carried to an explicit 'repudiation of middle-class life' in the belief that the simpler lives of the poor were fundamentally 'better than the lives of the sort of people I was brought up among'.[22] Few, of course, would have gone this far. But something of the new sympathy and understanding can be found even in the most formidable and influential of late nineteenth-century English social investigations, Charles Booth's *Life and Labour of the People in London* (1889–1903). Booth used a variety of methods, including participant observation, taking lodgings incognito in a working-class household, although for his main survey of poverty he did not use direct interviews, but relied on reports from school visitors. He took a great deal of oral evidence for his religious inquiry, but this was chiefly from clergy. For all its richness, his seventeen-volume masterpiece thus lacks the immediacy of working-class speech. Seebohm Rowntree, in developing Booth's method for his own study of York, *Poverty* (1901), did undertake direct interviewing, although his report was in the statistical tradition, avoiding quotation. But his later *Unemployment* (1911) uses direct quotations from interviewers' notes very effectively, and although this remains well short of Mayhew's standard, it provided an important early instance of the twentieth-century sociological survey, with its combination of tables and interview quotations. Another less well-known pioneering work is the cultural study of *The Equipment of the Workers* (1919), carried out by a high-minded adult education group at the St Philip's Settlement in Sheffield, using both a quantitative sample frame and a selected number of deeper qualitative interviews incorporating life histories. It is an odd book, but again an example of a method which might have been—although in the event was not—taken up at this time by historians.

A second line of influence from Booth's social survey leads more directly into history. One of his team of investigators was the young Beatrice Webb. Her contributions on dock labour and the sweated tailoring trade are the best industrial analyses in Booth's whole series. She also had early experience in door-to-door interrogation as a rent collector for Octavia Hill. Thus when she came to write her first independent historical study, *The Co-operative Movement in Britain* (1891), and later, with Sidney Webb, in their classic *History of Trade Unionism* (1894), she undertook the collection of oral, along with documentary, evidence in a highly systematic way. From the start, Beatrice combined searches through records with visits to Co-operative Societies and interviews with leading Co-operative personalities. Later she evolved with Sidney a method of occasional intensive field-work forays, setting up headquarters in lodgings in a provincial town for two or three weeks, and 'working hard; looking through minute books, interviewing and attending business meetings of trade unions'. Although at first Sidney preferred documentary work, being 'shy in cross-examining officials, who generally begin by being unwilling witnesses and need gentle but firm handling', they apparently hit upon a 'devastating technique of joint interview, in which they battered from either side the object of their attentions—sometimes a political opponent, sometimes an official who had not devoted much thought to the underlying implications of his official actions—with a steady left–right of question, argument, assertion, and contradiction, and left him converted, bewildered, or indignant, as the case might be'.[23] Later on Beatrice put these and other less dubious interviewing skills to effect in deliberately creating the evidence she wanted before the 1905 Poor Law Commission, both through procuring and briefing witnesses, and through her own cross-examining.

In their published histories, the Webbs cited only documentary sources. But they depended heavily on their interviewing both for their overall interpretation and for their treatment of facts. Each field-work visit resulted in an overall assessment of a particular organization, and a set of penetrating portraits of its personalities. The Webbs were careful to pass on their method to the school of British labour history of which they were the founders. Page Arnot, for example, followed it for his histories of the miners' trade unions. The notes on interviewing which Beatrice Webb published in *My Apprenticeship* (1926) still command respect. And it must surely be her example which inspired the leading economic historian, J. H. Clapham, in 1906, to call for the training of interviewers to collect 'the memories of businessmen' which were, in his view, 'the best original authorities' for recent economic history: and 'with them often die some of the most valuable records of nineteenth-century history'.[24] But nothing came of this; and, indeed, the Webbs themselves were to have few immediate followers, even in labour history.

It is no accident, however, that this innovative historical work by the Webbs was part of a lifetime dedicated to social change and practical politics. Where other notable experiments in the use of oral material by historians can be found in this period, the context is typically exceptional, and often literally at the frontiers. Thus as the British expanded their imperial control in Africa, missionaries and colonial civil servants would begin recording local native traditions; and by the 1900s especially in Uganda and among the Nigerian Yoruba, historical writing extensively drawn from their own oral tradition flourished among the conquered people themselves. In the Pacific the American missionary Sheldon Dibble organized his seminary class into a student research group, sending them out to 'the oldest and most knowing of the chiefs and people' armed with questions to elicit 'the main facts of Hawaiian history', for his *History of the Sandwich Islands* (1843). And more ambitiously, in the 1860s H. H. Bancroft, whose family firm were the largest booksellers, stationers, and publishers in the American Far West, decided to collect material on a very large scale for his historical studies of the recently colonized Pacific coast of California. Over a period of fifty years he employed altogether six hundred assistants, who built up, indexed, and abstracted his library. In addition to buying all the documents he could find and sending his agents to harass financially embarrassed families and corporations, he mobilized a whole army of reporters to extract conversations from surviving witnesses. Perhaps the most skilled of these was the Spanish-speaking Enrique Cerruti. Bancroft himself claimed that his library included 'two hundred volumes of original narratives from memory by as many early Californians, native and pioneer, written by themselves or taken down from their lips . . . There were a thousand, five thousand witnesses to the early history of this coast yet living, whom, as before intimated, Mr Bancroft resolved to see and question, all of them possible; and a thousand he did see, and a thousand his assistants saw, and wrote down from their own mouths the vivid narratives of their experiences'.[25] Bancroft's methods clearly had many weaknesses, and he proved unable to write up the material he collected in a convincing enough form. But in his willingness to use oral evidence, he set a precedent which was subsequently followed both in serious scholarship and in popular local journalism. Frederick Jackson Turner partly reached his famous thesis on the significance of the open frontier in this way. Similarly, from the 1920s it was a regular policy of the Arizona *Republican* to collect stories for publication from 'old-timers' at annually organized pioneers' reunions. And certainly Bancroft himself had been able, through his own private wealth, to organize one of the most elaborate purely historical research enterprises of the nineteenth century, anticipating some of the giant public and privately funded projects of a hundred years later.

It may perhaps be a salutary warning that although his library now forms the centre of the great Berkeley university campus, as a historian Bancroft is now largely forgotten. In this he stands in sharp contrast to another pioneer of oral history, the French historian Jules Michelet. Michelet is rightly remembered; and more needs to be said at this point about his use of oral evidence. He is a remarkable figure: both the leading professional of his age, and a great popular historian; and as imaginative in seeing the possibilities of documentary archives, as of oral tradition. Besides this he was one of the first historians to bring an understanding of the land and landscape into his writing. His influence was diffuse. One can see it in W. G. Hoskins, following *The Making of the English Landscape* (1955) along the hedgerows; or in France, the great medievalist Marc Bloch combining his searches in archives with the study of field patterns, place names, and folklore, tramping round the French countryside talking with a peasantry who in the early twentieth century still worked the land with some of the means and spirit of their medieval predecessors. Michelet himself used oral evidence, especially in his *History of the French Revolution*, where he realized that the official documents preserved only one side of the political story. In 1846 he had also published *Le Peuple*, a remarkable essay on the impact of mechanization on the social classes of France. Its preface contains a striking—indeed passionate—statement of how he came to his method, and gained from it. He had been collecting information outside Paris for ten years, starting with Lyons, and then moving to other provincial towns, and into the countryside. 'My inquiry among *living* documents', he wrote, 'taught me many things that are not in our statistics . . . The mass of new information I have thus acquired, and which is not in any book, would scarcely be credited.' This was how he had first noticed the immense increase in the use of linen articles by poor families, and from this deduced an important shift within the structure of the family itself:

This fact, important in itself as an advance in cleanliness . . . proves an increasing stability in households and families—above all the influence of woman, who, gaining little by her own means, can only make this outlay by appropriating part of the wages of the husband. Woman, in these households, is economy, order, and providence . . . This was a useful indication of the insufficiency of the documents gathered from statistics and other works of political economy, for comprehending the people; such documents offer partial, artificial results, views taken at a sharp angle, which may be wrongly interpreted.

Michelet felt exceptionally at ease with this kind of research. This was partly because of his early life in a Parisian printer's family. Interviewing brought him back close to his own social origins, from which he had been separated through his education. 'I have made this book of *myself*, of my life, and of my heart. It is the fruit of my experience . . . I have derived it from my own observation, and my intercourse with friends and neighbours; I have gleaned it from the highway.' He seems to have been considerably happier talking to poor people than he was with the social class into which he had risen:

Next to the conversation of men of genius and profound erudition, that of the people is certainly the most instructive. If one be not able to converse with Béranger, Lamennais, or Lamartine, we must go into the fields and chat with a peasant. What is to be learnt from the middle class? As to the *salons*, I never left them without finding my heart shrunk and chilled . . .

Even so, it had been far from easy for Michelet to reach an open recognition of such feeling. As a young man, competitive, moving upwards through education, he had become intensely withdrawn. 'The fierce trial at college had altered my character—had made me reserved and close, shy and distrustful . . . I desired less and less the society of men.' His rediscovery of others and of himself came through his teaching at the École Normale:

Those young people, amiable and confiding, who believed in me, reconciled me to mankind . . .

The lonely writer plunged again into the crowd, listened to their noise, and noted their words. They were perfectly the same people . . .

(My pupils) had done me, without knowing it, an immense service. If I had, as a historian, any special merit to sustain me on a level with my illustrious predecessors, I should owe it to teaching, which for me was friendship. Those great historians have been brilliant, judicious, and profound; as for me, I have loved more.

Nineteenth-century historians were not given to self-analysis. Michelet therefore provides, in the few, vivid paragraphs of this preface, a powerful indication of an increasing barrier to the practice of oral history: class. The nineteenth century was everywhere an age of increasing class and status consciousness. Historians were themselves evolving into a close profession, recruited through education. The very few who made their way into it from relatively humble backgrounds were much more likely to remain, because of the difficult experience of social mobility, withdrawn, like Michelet in his early adulthood. Among these Michelet was exceptional: few shared either the political commitment or the personality which enabled him to break back into easy contact with the people. As we shall see, the exclusive professionalism exemplified in Germany proved more compelling. And the very fecundity of production of secondary oral sources made it more possible, by the mid-nineteenth century, for a great historian to write without the use of any 'living documents'.

Michelet himself knew this as well as any man of his time. In 1831 he had been appointed chief of the historical section of the

National Archives of France, an immense collection which had been brought together when the French Revolution 'emptied the contents of monasteries, castles, and other receptacles on one common floor'. He used it for his own *History of France* (1833–67), and his afterword to its second volume provides an equally telling psychological insight, this time into the personality of the archival historian. It is a species of fantasy-hymn:

The day will be ours, for we are death. All gravitates to us, and every revolution turns to our profit. Sooner or later, conquering or conquered come to us. We have the monarchy, safe and sound, from its alpha to its omega . . . the keys of the Bastille, the minute of the declaration of the rights of man . . .

As for me, when I first entered these catacombs of manuscripts, this wonderful necropolis of national monuments, I would willingly have exclaimed . . . 'This is my rest for ever; here will I dwell, for I have desired it!'

However, I was not slow to discern in the midst of the apparent silence of these galleries, a movement and a murmur which were not those of death. These papers and parchments, so long deserted, desired no better than to be restored to the light of day: yet they are not papers, but lives of men, of provinces, and of nations . . . All lived and spoke, and surrounded the author with an army speaking a hundred tongues . . .

As I breathed on their dust, I saw them rise up. They raised from the sepulchre, one the head, the other the hand, as in the Last Judgement of Michelangelo, or in the Dance of Death. This galvanic dance, which they performed around me, I have essayed to reproduce in this work.

The notion that the document is not mere paper, but reality, is here converted into a macabre gothic delusion, a romantic nightmare. But it is nevertheless one of the psychological assumptions which underpin the documentary empirical tradition in history generally, and not in France alone. In a much more careful, veiled form, for example, one may find the same dream in that early masterpiece of English professional scholarship, F. W. Maitland's *Domesday Book and Beyond* (1897). 'If English history is to be understood, the law of Domesday Book must be mastered.' Maitland looks forward to a future in which the documents have all been reorganized, edited, analysed. Only then, he writes, 'by slow degrees the thoughts of our forefathers, their common thoughts about common things, will have become thinkable once more . . .'. And the dream is there in the title itself. 'Domesday Book appears to me, not indeed as the known, but as the knowable. Beyond is still very dark: but the way to it lies through the Norman record.'[26]

It was this documentary tradition which emerged during the nineteenth century as the central discipline of a new professional history. Its roots go back to the negative scepticism of the Enlightenment as well as to the archival dreams of the Romantics. We have already met the Scottish historian William Robertson at breakfast with Dr Johnson. Robertson, in his *History of the Reign of Charles V* (1769), publicly reprimanded Voltaire for his failure to cite sources. He had himself gone to unusual lengths to base his *History of Scotland* (1759) on original documents, and was able to cite seven major archives, including the British Museum, although 'that Noble Collection' was 'not yet open to the public. . . . Publick archives, as well as the repositories of private men, have been ransacked . . . But many important papers have escaped the notice of (others). . . It was my duty to search for these, and I found this unpleasant task attended with considerable utility . . . By consulting them, I have been enabled, in many instances, to correct the inaccuracies of former Historians.' Archival research at this stage is thus seen essentially as a distasteful corrective duty, rather than a creative skill. And it is the same negative scepticism which leads Robertson to reject out of hand the entire oral tradition of early Scottish history, dismissing it as 'the fabulous tales of . . . ignorant Chroniclers'. The history of Scotland before the tenth century was not even worth study. 'Everything beyond that short period to which well attested annals reach, is obscure . . . the region of pure fable and conjecture, and ought to be totally neglected.'[27]

It is less easy to see why this sceptical approach should have triumphed in the nineteenth century. Paradoxically, the same romanticism which breathed life into the documentary method also set going folklore collecting all over Europe, and recovered for the great epics and sagas of oral traditions the respect which they deserved. In Britain the folklore movement developed independently of professional history, on a local antiquarian or literary basis, largely amateur, and adopted its own specia evolutionary theory of 'survivals' from Darwin. In France and Italy—where interest could be traced back at least to the eighteenth-century philosopher-historian Vico—folklore became a much more respected branch of scholarship. But it gained its greatest hold in Scandinavia and in Germany. Here, as in Britain, there had been earlier instances of collecting and publishing, but this initial antiquarianism was succeeded by the sophisticated methodology of ethnology, using a historical-geographical framework for systematic documentation and comparison. In this form it has, as we shall see, made a direct contribution to the modern oral history movement. At the same time it came to be seen as an important way of recovering a lost national spirit and culture, not only in Scandinavia, but also in Germany.

Equally important, the Romantic Movement led in the philosophy of history to a widespread acceptance of the importance of cultural history and the need to understand the different standards of judgement of earlier epochs and, eventually, other societies. This was again especially true of Germany where the narrowly confident universalistic rationalism of the Enlightenment had been resisted almost from the start, most notably by Herder, with his belief that the very essence of history was in its plenitude and variety. Here already were the first steps towards a cultural relativism. And it was from Vienna that, at the end of the nineteenth century, the modern understanding of individual personality through psychology originated, carrying with it the implications of a less judgemental, more relativist attitude towards individuals in history. German philosophers of history unfortunately took little consistent interest in psychology. But the possibility of a new understanding of the historical value of individual life stories was certainly there, and at least one German philosopher, Wilhelm Dilthey, came at times very close to it, as is demonstrated in some of his reflections on the meaning of History:

Autobiography is the highest and most instructive form in which the understanding of life confronts us. Here is the outward, phenomenal course of a life which forms the basis for understanding what has produced it within a certain environment . . .

The person who seeks the connecting threads in the history of his life has already, from different points of view, created a coherence in that life which he is now putting into words . . . He has, in his memory, singled out and accentuated the moments which he experienced as significant; others he has allowed to sink into forgetfulness . . .

Thus, the first problem of grasping and presenting historical connections is already half solved by life.[28]

How was this opportunity lost? What led the documentary method to its narrowing, scarcely mitigated triumph in the very same decades through German example? This is a question which needs to be more fully explored. But part of the explanation undoubtedly lies in the changing social position of the historian. The development of an academic, historical profession in the nineteenth century brought with it a more precise and conscious social standing. It also required that historians, like other professionals, should have some form of distinctive training. And both the research doctorate, and the systematic teaching of historical methodology, are derived from Germany. Research training was begun by Leopold von Ranke after his appointment in 1825 as professor at Berlin. Ranke was already thirty, but he was to live to the age of ninety, and during the succeeding decades his research seminar became the most influential historical training-ground in Europe. He was in some ways an old-fashioned figure, a sceptic as much as a romantic despite his fascination with medieval Germany. It was a rejection of Scott's novels as factually unreliable which first led him to resolve that in his own work he would avoid all fabrication and fiction, and stick severely to the facts. But in his first great masterpiece, the *Histories of the Latin and Germanic Nations* (1824), despite his famous destruction of Guiccardini's credibility and his dictum that history should be written *wie es eigentlich gewesen ist* (as it really was), he also declared himself opposed to research for its own sake; it was only in the final stage of his work that he had resorted to archives for confirmation. And although the *History of the Popes* (1837) was based on a more active approach, he certainly never shared the positive fascination with archives of his contemporary Michelet. Indeed, later in life he evolved a routine which avoided any direct contact with archives. Documents were brought to him in his own home by his own research assistants, who would read them aloud. If he so instructed, the assistant would make a copy of the document. Ranke would work each day from 9.30 am until 2 pm with his first assistant, and from 7 pm with his second, in between taking a walk with a servant in the park, dinner, and a brief sleep. What mattered most was the relentlessness of his systematic, critical spirit. He directly trained more than a hundred eminent German university historians. In his research seminar, although they were allowed to choose their own topic, he set them on to medieval documentary work simply because that was the most difficult to master. And when professional training began to spread, first to France in the 1860s and later elsewhere in Europe and in America, it was founded on Ranke's assumptions. C.-V. Langlois and Charles Seignobos of the Sorbonne opened their classic manual, *Introduction to the Study of History* (1898), with the unqualified statement: 'The historian works with documents. . . . There is no substitute for documents: no documents, no history.'[29]

The documentary method not only provided an ideal training-ground, but it offered three other key advantages to the professional historian. First, the test of a young scholar's ability could become the writing of a monograph, the exploration of a corner of the past, perhaps minute, but based on original documents, and therefore, in that sense at least, original. Secondly, it gave to the discipline a distinct method of its own, which—unlike the use of oral evidence—could be claimed as an expert specialism, not shared by others. This self-identification around a distinct method—like the archaeological dig, the sociological survey, the anthropologist's field-trip—is typical of nineteenth-century professionalism and had the added function of making the evaluation of expertise an internal matter, not subject to the judgement of outsiders. Thirdly, for the increasing number of historians who preferred being shut up in their studies to mixing with either the society of the rich and powerful or with ordinary people, documentary research was an invaluable social protection. By cutting themselves off they could also pretend to an objective neutrality, and thence even come to believe that insulation from the social world was a positive professional virtue. Nor is it accidental that the cradle of this academic professionalism should have been nineteenth-century Germany, where university professors

constituted a narrow patrician middle-class group, particularly
sharply cut off through their isolation in small provincial
towns, political impotence, and the acute hierarchical status-
consciousness of Germany, from the realities of political and
social life.

In Britain the full development of these tendencies came relatively late. Although the documents of the Constitution had
been firmly enough enshrined by Bishop Stubbs, eminent late
nineteenth-century scholars like Thorold Rogers and J. R. Green
did not trouble to footnote their main works, and even the
*Cambridge Modern History*, launched by Lord Acton in 1902 as
'the final stage in the conditions of historical learning', was
intended to be without footnotes.[30] The academic establishment
was still widely linked both through kin and personal careers with
London society and the political world. Thus Beatrice and Sidney
Webb, in the midst of their political work for the Poor Law
Commission, were also writing the chapter on social movements
for the *Cambridge Modern History*; while R. C. K. Ensor, who
wrote the highly successful Oxford volume on *England 1870–
1914* (1936), had spent most of his life in journalism, politics, and
social work. Lewis Namier's famous puncturing of the old Whig
school of history, *The Structure of Politics at the Accession of
George III*, only came out in 1929. It was not until the post-Second
World War expansion of the universities that the research
doctorate became the standard method of entry into the historical
profession. Its full advantages, and disadvantages, are therefore
a comparative novelty to British historians.

By this stage the ideal moment of the documentary method had
already passed. It always had its critics. Even Langlois and
Seignobos warned against the 'mental deformations' which critical scholarship had led to in Germany: a textual criticism lost in
insignificant minutiae, separated by a chasm from not just general
culture, but the larger questions of history itself. 'Some of the
most accomplished critics merely make a trade of their skill, and
have never reflected on the ends to which their art is a means.'
They also commented on the ease with which a 'spontaneous
credulity' of anything documented can develop (rather characteristically instancing memoirs as a type of document deserving

'special distrust') and argued for both analytic criticism and comparative evidence for establishing facts: 'It is by combining observations that every science is built up: a scientific fact is a centre on which several different observations converge.' Their first point is repeated by R. G. Collingwood in *The Idea of History* (1946), who condemns a training that 'led to the corollary that nothing was a legitimate problem for history unless it was either a microscopic problem, or else capable of being treated as a group of microscopic problems'; he instances Mommsen, who 'was able to compile a corpus of inscriptions or a handbook of Roman Constitutional law with almost incredible accuracy . . . but his attempt to write a history of Rome broke down exactly at the point where his own contributions to Roman history began to be important'.[31] If such comments had force then they have still more today in a rapidly changing world which demands explanations for its own instability. An escape from major problems of historical interpretation into microscopic investigation is increasingly difficult to justify. The documentary tradition has thus found itself increasingly on the defensive in the face of the growth of the social sciences, with their claims to superior powers of interpretation and theory.

Still more critically, the documentary school faces a shifting of its very foundation, for the document itself has changed its social function in two ways. First, the most important communications between people are no longer made through documents (if they ever were) but orally, by meeting or telephone. Secondly, the record has lost its innocence (if it ever had one); it is now understood to have potential value as future propaganda.

The stages of this change have been shrewdly discussed by A. J. P. Taylor, the prime master of the modern English documentary school. They first presented themselves in the documentation of diplomatic history:

The historian of the Middle Ages, who looks down on the 'contemporary' historian, is inclined to forget that his prized sources are an accidental collection, which have survived the ravages of time and which the archivist allows him to see. All sources are suspect; and there is no reason why the diplomatic historian should be less critical than his colleagues. Our sources are primarily the records which foreign offices keep of their main dealings with each other; and the writer who bases himself solely on the archives is likely to claim scholarly virtue. But foreign policy has to be defined as well as executed . . . Public opinion had to be considered; the public had to be educated . . . Foreign policy had to be justified both before and after it was made. The historian will never forget that the material thus provided was devised for purposes of advocacy, not as a contribution to pure scholarship; but he would be foolish if he rejected it as worthless . . . The same is true of the volumes of memoirs, in which statesmen seek to justify themselves in the eyes of their fellow country-men or of posterity. All politicians have selective memories; and this is most true of politicians who originally practised as historians. The diplomatic record is itself drawn on as an engine of publicity. Here Great Britain led the way . . .

in the parliamentary Blue books; followed in the 1860s by France and Austria, and later by Germany and Russia. Specially favoured historians were also allowed access to the archives to write their histories. Next came the fuller publications from archives by governments, normally either to justify or to discredit their predecessors. The first of these great collections was the French series on the origins of the 1870 War, published from 1910 onwards; but 'the real battle of diplomatic documents' opened at the end of the First World War with the Russian publication of the secret treaties, and then successive series issued from Germany, France, Britain, and Italy.[32]

From the 1920s, therefore, no diplomat could possibly forget that any document which he eventually retained might later be used against him. The original record must therefore be as judicious as possible; and periodic weeding of the files was always desirable. Meanwhile a similar process of change had started with home documents. Confidential cabinet papers were being kept by politicians and some were able to use them in their memoirs. For a long time this tendency was fought, but effective recognition that no document could be regarded as permanently confidential (except perhaps by the police or secret services) came with the reduction of the waiting period for normal access to scholars to a mere thirty years. The consequence can be seen in the comment made to A.J.P. Taylor by Richard Crossman, former Cabinet minister: 'I've discovered, having read all the Cabinet papers about the meetings I attended, that the documents often bear virtually no relation to what actually happened. I know now that the Cabinet Minutes are written by Burke Trend (secretary to the Cabinet), not to say what *did* happen in the Cabinet, but what the Civil Service wishes it to be believed happened, so that a clear directive can be given.' In the decades before the First World War, however, such tampering was only beginning. Equally important was the fact that this was the golden era of the personal letter. When dealing with the post-First World War period, Taylor himself has argued for the use of 'non-literary sources . . . The more evidence we have, the more questioning we often become. Now we have recording instruments for both sight and sound.' But he saw such needs in contrast to an earlier period:

The seventy years covered by this book are an ideal field for the diplomatic historian. Full records were kept, without thought that they would ever be published, except for the occasional dispatch which a British statesman composed 'for the Blue Book'. It was the great age of writing. Even close colleagues wrote to each other, sometimes two or three times a day. Bismarck did all his thinking on paper, and he was not alone. Only Napoleon III kept his secrets to himself and thwarted posterity. Now the telephone and the personal meeting leave gaps in our knowledge which can never be filled. While diplomacy has become more formal, the real process of decision escapes us.[33]

We have arrived, in short, at the age of the telephone and the tape recorder: a change in methods of communication which will in time bring about as important an alteration to the character of history as the manuscript, the printing press, and the archive have in the past.

It looks, too, as if it may be a swifter change. The technological basis has certainly evolved with great rapidity. The first recording machine, the phonograph, was invented in 1877, and the steel wire recorder just before 1900. By the 1930s a considerably improved version was good enough for use in broadcasting. A decade later magnetic tape was available and the first tape recorders of the reel-to-reel type sold on the market. The much cheaper cassette recorders came in the early 1960s. Today it is practicable for any historian to consider using a recorder in collecting evidence. This transformation of technology provides one reason why the modern oral history movement has its origins in most countries in substantially and quite often nationally funded enterprises, yet has more recently been growing equally fast as a form of diffused local and popular history.

Let us turn then to the pattern of the revival, bearing in mind the constraints imposed by resources. Where has oral history grown most strongly? How have the intellectual contributions to the reviving use of oral evidence, and the patterns of sponsorship, varied from place to place? We can start most conveniently with North America, which has seen the most explosive growth of all.

The antecedents of the movement there go back many years. H. H. Bancroft's interviewing of the 1860s was succeeded by other intermittent work on the frontier settlements; and the American Folklore Society dates back to 1888. More important was the great break forward of American urban sociology from its English-influenced origins to the Chicago studies of the 1920s, like Harvey Zorbaugh's *Gold Coast and Slum* (1929), vibrant with direct observation and interpretation of city life, and centrally concerned with documenting and explaining it. In these early years the Chicago sociologists were remarkably inventive in their methods, making use of direct interviewing, participant observation, documentary research, mapping, and statistics. They developed a special interest in the life history method for studying two aspects of urban social problems.

The first was a practical contribution to criminology. Clifford Shaw's masterpieces, such as *The Jack Roller: a Delinquent Boy's Own Story* (1930) or *Brothers in Crime* (1938), used a mere few of many hundreds of life stories which he collected from the youth of Chicago's inner-city slums. Shaw's technique can be traced back not only to Henry Mayhew's lives of London criminals, but also to the traditional seeking of confessions from convicts on the scaffold or—as the reformers renamed the prison—in the penitentiary. In Britain John Clay, the prison chaplain at Preston, encouraged inmates to write or dictate 'short narratives of their lives, their delinquencies, their self-convictions, and their penitence', believing such stories to illustrate 'a history of which we are yet too ignorant,—the actual social and moral state of our poor fellow-subjects'. Clay published some of the stories he gathered in his prison reports from the 1840s, using them to argue in support of the separate cell system. Similarly in America in the 1900s Judge Ben Lindsay of Denver used 'life speech' confessions as a means of treating youths in his model juvenile court, and Dr William Healey, founder of the Institute for Juvenile Research, which Shaw later led, and originator of the psychiatric case conference, used a parallel 'own story' technique both for therapy and for seeking understanding of delinquents' own attitudes. The crucial influence of this life story approach in social casework and therapy today is so fundamental that it is taken for granted, but was then new. Equally Shaw's books, setting life stories with great care in their family and social context, showed so convincingly that delinquency was not just the outcome of pathological character but a response to social deprivation, that eventually they seemed redundant: the point was taken.[34]

The second focus, long-term social change, overlaps more obviously with oral history, drawing on older informants: but as much by persuading them to write autobiographies or diaries, or lend letters, as by life story interviewing. Thus W. I. Thomas and F. Znaniecki in their massive pioneering account of immigration, *The Polish Peasant in Europe and America* (1918–20), gave an entire volume to the *Life Record of an Immigrant*, a specially solicited written autobiography which provides a link between studies on social disorganization in Poland and the origins of emigration, and on the Polish community in Chicago. Znaniecki continued to work in both Poland and America. He founded the distinctive 'humanistic tradition' in Polish sociology, which systematically uses public competitions to collect written 'memoirs' on particular themes. It was developed by radical social commentators who used it to demonstrate the plight of the Polish peasantry and unemployed in the 1930s (and inspired a similar British volume of *Memoirs of the Unemployed* collected through a radio appeal in 1933).[35] In post-war Poland memoir competitions have become an astonishingly lively form of popular culture. A continuation of interest can be seen in John Dollard's early study, *Criteria for Life History* (1935). But direct links with more recent life story sociology are surprisingly rare. Polish work is little known in the West; while the Chicago school, despite such

a promising beginning, before long became a victim of profes-
sionalization among sociologists, and retreated from the immediacy of the city around it to the security of research doctorates
based on statistical analysis and abstract general theory.

Its legacy was not forgotten. It is still alive in the work of the
Chicago journalist and oral historian, Studs Terkel. Shaw's own
Institute launched a fertile revival of life stories in the sociology
of deviance with the publication of *The Fantastic Lodge: the
Autobiography of a Girl Drug Addict* (1961) from recordings
made with Howard Becker. Another link with the present is
through American anthropology. The inter-war years were a
period in which the general tendency in anthropology was
strongly influenced by Malinowski's argument that oral traditions, just because their key function was to justify and explain
the present, had virtually no value as history: myth was 'neither a
fictitious story, nor an account of a dead past; it is a statement of
a bigger reality still partially alive'. Although his views applied
more to oral tradition than to direct personal life story evidence,
they undoubtedly inhibited any move in this direction too. The
European anthropologists who had scattered for their field-work
to the remotest corners of the colonial empires rarely showed
concern for the actual words of their informants. In America,
however, anthropologists working among North American
Indians and in Mexico were also in contact with the development
of psychology and sociology, and took up the life-history
method. Thus the work of Oscar Lewis and Sidney Mintz from
the 1950s can be traced back, through Leo Simmons's *Sun Chief*
(1942), an oral history project jointly sponsored by the
anthropologists, psychiatrists, and sociologists at Yale, to Paul
Radin's *Crashing Thunder* (1926), an American-Indian life story
inspired by the need 'of obtaining an inside view of their culture
from their own lips and by their own initiative'. These antecedents include Ruth Landes's brilliant portrayal of nomadic
Canadian Indian hunting people in *The Ojibwa Woman* (1938),
which includes a rare early collection of women's life stories.[36]

Most striking of all was an experiment launched under government sponsorship to fight unemployment in the New Deal: the
Federal Writers' Project of the 1930s. An astonishing series of

life story interviews was collected right across the country with former black slaves, workers, and homesteaders, the richness of which is only now beginning to be fully appreciated. Much of this material remained unpublished, but one contemporary selection, published in North Carolina and edited by W. T. Couch under the title *These Are Our Lives* (1939), shows a remarkable understanding of the radical potential of oral history. Sociology, Couch argued, had been 'content in the main to treat human beings as abstractions', or when case histories were used, to dissect them as 'segments of experience' in the analysis of particular problems such as social maladjustment. But it would be possible, 'through life histories selected to represent the different types present among the people'—in appropriate proportions—to portray an entire community. His own collection of life histories was intended to represent for their region 'a fair picture of the structure and working of society. So far as I know, this method of portraying the quality of life of a people, of revealing the real workings of institutions, customs, habits, has never before been used for the people of any region or country.'[37]

Despite such anticipations, it was from another direction that he key step in the modern movement came: political history. Oral history', the (American) Oral History Association declared, 'was established in 1948 as a modern technique for historical documentation when Columbia University historian Allan Nevins began recording the memoirs of persons significant n American life'. The Columbia approach, the privately financed 'great man' recording project, proved immensely attractive o both national foundations and local fund-givers, and especially to retiring politicians. Indeed, for at least two decades it was 'oral history' in America—and only from the 1970s was the method vigorously revived for Indian history, black history, and folklore, and extended into new fields like women's history. The North American scene is now one of both variety and vitality, specially if one includes the Canadian Oral History Association formed in 1974, with its own journal. The Oral History Association itself has 1,500 members and its 1982 directory listed 500 current projects. In the two countries together there were already by 1971 100,000 recorded hours of interviews collected, and over a million pages of transcript. These figures reflect the sheer resources which have made such a scale of field-work possible. One consequence is that an exceptionally high proportion of American oral historians are archivists. But alongside community work there is also a growing academic current, clearly reflected in the papers and bibliographies of the Association's *Oral History Review* in the 1980s; and especially through the influence of Ron Grele, who now heads the Columbia programme, founder of the *International Journal of Oral History* in 1980, in introducing perspectives from anthropology and from European oral history, it has become more reflective and outward-looking.

The second great concentration is in western Europe. It is here, beginning in Bologna in 1976 and Colchester in 1979, that the biennial international oral history conferences are held. But as they show, there is considerable activity elsewhere, too. In Latin America, the lively and varied scene draws on several influences: the life story school in American anthropology; Mexico's impressive national oral history programme, recording social movements, politics, and culture, developed since 1959; the social mobility research of the Argentinian sociologists, Jorge Balan and Elizabeth Jelin, who analysed life stories statistically on a sample basis in their Monterrey study of the 1960s, again in Mexico; and contemporary political history programmes as at Rio de Janeiro in Brazil, which have an added urgency in a continent where repeated political upheavals regularly destroy written documentation. In Australia oral history has its own association and journal, and has brought local and labour historians together with anthropologists of the Aboriginal peoples. In Asia, Singapore has its official archive, in striking contrast to India where most work still depends on British scholars and broadcasters. In the Communist world there has been very little tape-recorded oral history, but in addition to Poland with its popular autobiographical competitions and the after-hours factory memoir groups, there has been a fluctuating interest in oral history in China. The national collecting of revolutionary memories began in the 1950s, quite soon after the Communist victory; in 1958 with the first wave of the Cultural Revolution, the emphasis shifted to grass-roots factory, brigade, and village history groups, including illiterate older workers, investigating the 'tortuous and complex', 'hard and glorious struggle of the working classes'; but both campaigns disappeared in the anti-intellectualism and ultimate chaos of the Cultural Revolution years.[38] In their aftermath it was left to an American, William Hinton, to return and make sense of them in a masterpiece of oral history, *Shenfan: the continuing revolution in a Chinese village* (1983).

In Africa, and in Israel too, American and European influences have combined in different ways with resurgent nationalism. In the post-colonial era the history of Africa, which had been that of the imperial powers, abruptly shifted its focus to the largely undocumented African nations. But ironically the new school which arose, making increasingly subtle use of oral tradition as a source, has been predominantly Anglo-American; and excepting some very recent projects in South Africa, includes little social history recording the experience of ordinary African people. For Israel, by contrast, after the systematic destruction of Jewish communities under Fascism, oral evidence from witnesses of every variety has become a vital part of a national and cultural struggle for survival.

Within Europe there are parallels, more remotely in the link between nineteenth-century nationalism and folklore collecting, but also directly. In Italy one of the origins of contemporary oral history has been the network of local centres for studying the anti-Fascist wartime Partisans. Subsequently the perplexing political and social results of the post-war boom, with peasant immigration into the cities and changing working-class consciousness, created an interdisciplinary oral history fashion in the 1970s, as well as stimulating sustained research: notably by the sociologist Franco Ferrarotti on the slums and shantytowns of Rome, Sandro Portelli's cultural interpretations of the steelmakers of Terni, and the cluster of social history studies of peasants, workers, and women in Piedmont and Turin. It is from his last circle that Italy's oral history journal, *Fonti orali*, has been edited by Luisa Passerini; and it includes Nuto Revelli, most widely read of Italian oral historians, whose powerful books of testimonies have moved, indicatively, from war and resistance to peasant poverty and, finally, to the memories of mountain peasant women.[39] Documenting Fascism was also a principal object in the Netherlands where since 1962 oral history has grown from a well-organized co-operation between contemporary political historians, the International Institute for Social History, and Dutch radio, subsequently broadening into social history.

Conversely, the development of oral history in Spain had to wait for the ending of Franco's long regime, with the path led by the English oral historian Ronald Fraser. The late start of an oral history movement in Germany is again explained by the impact of Nazism, which discredited the folklore movement by espousing it, and at the same time destroyed the germs of a more fruitful approach to survey research which by the early 1930s had been shown in a study like *Marienthal*. More important, Nazism left a generation ashamed of its own experience, and a nation anxious to bury its past rather than to investigate it. Nevertheless, by the 1980s the social history research on the Ruhr working class led by Lutz Niethammer stood between both a growing range of local history projects and also an organized network of theoretically self-conscious life story sociologists.

It is much less clear why France, with a widespread interest in war and resistance history, and with the example not only of Michelet but of the sociological school of Durkheim (which drew together anthropological and folklore material), and even a remarkable pioneering work by Maurice Halbwachs on the social nature of memory to build upon, was also late in developing activity in oral history. The extreme concentration of French academic research in specialized institutes in Paris, lacking links with local communities, may be one reason. Even today most French oral history community work is to be found outside France, in Belgium; while the pioneering research came from the sociologists Daniel and Isabelle Bertaux, and from Philippe Joutard and his interdisciplinary group of linguists, ethnologists, and historians in the south at Aix-en-Provence.[40]

It is, however, in Scandinavia, and in the British Isles, that the most strongly established European developments can be found. In Scandinavia the roots lie in the systematic folklore collecting of the nineteenth century. The first archives for direct field-work were set up in Finland as early as the 1830s. The Finnish example was followed especially in Sweden. Students at the University of Uppsala formed dialect societies in the 1870s to collect provincial words and expressions which they feared were threatened with extinction. Already by the 1890s this collecting had been systematized into a national questionnaire interview survey, answered in a thousand different locations over the whole country, and by 1914 the Institute for Dialect and Folklore Research was founded with financial support from the Swedish Parliament. The scope of its collecting gradually widened into a national study of rural society, culture, and economy. And from 1935 the Institute made regular use of recording machines in its field-work—probably the first organization to do so for the purposes of historical research. Closely linked to this field-work was the special development of ethnology in Scandinavia as a central academic discipline in the social sciences, fusing social history and sociology. In the 1950s, led by the Norwegian historian Edvard Bull, ethnological field-work collection was extended to the urban and industrial populations; and by the 1970s ethnologists such as Orvar Löfgren and Sven Ek were using this earlier work to study long-term social change. There have also been notable experiments in popular history, through the imaginative museum and broadcasting services, and also the workers' factory history campaign launched by the Swedish writer Sven Lindqvist in his challenging book, *Grav där du står* (1978) and the tented travelling exhibition under the same title—'Dig Where You Stand'.

This early Swedish example proved of particular importance in the development of oral history in Britain. Here again a strong interest had developed in folklore, mainly on an amateur basis. But in Ireland and Wales, and to a lesser extent in Scotland, this was reinforced through association with nationalist movements. The Irish Government began to assist collecting in 1930, and in 1935 set up the Irish Folklore Institute. From the start this had direct links with Swedish scholars and also made use of recording machines. In Wales, the main centre became the Welsh Folk Museum at St Fagans; in Scotland systematic collecting was led from the Edinburgh University School of Scottish Studies, whose archive was started in 1951, originally with a Gaelic and literary focus but before long also drawing in social and English language material. Lastly, in England the major comparable enterprises were of the Dialect Survey begun about 1950 from Leeds University and the subsequent Centre for English Cultural Tradition and Language at Sheffield.

It was post-war political change, however, which in Britain too lay behind the revival of oral history. As colonial Africa moved to independence its new nations needed a history of their own. From the 1950s, led by the Belgian scholar Jan Vansina—later from Wisconsin—and John Fage and Roland Oliver from Britain, historians began to collect their own oral material in the field, alongside anthropologists, exchanging experience of methods and interpretation with them.

The coming to power of the working-class movement in the 1945 Labour government and the popular confidence from the long post-war boom years brought, more slowly, a parallel change at home: a quickening interest in labour history, by the 1960s broadening into social history, paralleled by a new enthusiasm for working-class autobiography and later for television series using ordinary people's memories like 'Yesterday's Witness'. Some historians also became aware through their own radio activities of the remarkable resources of the BBC Sound Archives, which had been founded in the 1930s. The crucial influence came, however, through a new sociology of the 1950s concerned, not just with poverty, but with working-class culture and community in its own right. Some of these classic studies, such as Peter Townsend's *The Family Life of Old People* (1957) and Brian Jackson's and Dennis Marsden's *Education and the Working Class* (1962), made an effective use of individual working-class memories, while Richard Hoggart's semi-autobiographical *The Uses of Literacy* (1957) sought to interpret working-class forms of thought in speech and 'oral tradition'. With Edward Thompson's *The Making of the English Working Class* (1963) this new sympathy was matched with a history which sought 'to rescue the poor stockinger, the Luddite cropper, the "obsolete" hand-loom weaver, the "utopian" artisan, and even the deluded follower of Joanna Southcott, from the enormous condescension of posterity', seeing their ideas instead as 'valid in terms of their own experience'.[41]

This convergence of sociology and history was encouraged through the founding of the new universities of the 1960s with their interdisciplinary experiments, and the rapid expansion of a sociology which was showing an increasing concern with the historical dimension in social analysis. The potential of oral history was brought home through the popular success of Ronald Blythe's *Akenfield: Portrait of an English Village* (1969), a blend of literature, history, and sociology based on tape recordings from Suffolk country people. Nor is it an accident that one of the most significant later books using oral evidence is a historical study of the relationship between religion, economics, and class-consciousness by a sociologist—Robert Moore's *Pit-Men, Preachers and Politics* (1974). Thea Vigne and I started our own national interview survey of family life, work, and community before 1918 from the sociology department of Essex University in 1968. We drew on the sociological experience of colleagues such as Peter Townsend and were given the financial support of the newly established Social Science Research Council.

Since then oral history has grown fast in Britain. The Oral History Society was formed in 1973, and within six years had some 600 members. It drew on each of these developing strands, and others too. The larger new projects tended to be in social history, funded by the government Research Council and owing something to sociological influence. But others were started in colonial and military history. There was also renewed activity in those branches of history which, for different reasons, had retained at least a minority tradition of oral field-work: recent political history, labour history, and local history.

In recent political history the change has been least obvious, because although often not cited, there has been continuous use of the interview as a method of exploration, discovering documents, and checking interpretation. A modern political biographer would always seek to learn from conversation with a subject, just as, for example, John Morley did from the ageing Gladstone. David Butler could even write that his *Electoral System in Britain 1918–51* (1953) 'owes more to the personal recollections of the surviving protagonists than to any published lives or historians'.[42] But the advent of the tape recorder provided a more systematic method of collecting interview evidence.

With labour history the line of development from the Webbs is clearer. There is now a great deal of activity in this field, including substantial projects; and oral evidence has from the start been one of the distinctive marks of the *History Workshop* movement, which began out of working-class labour and social history at Ruskin College, Oxford, and has widened its range to address itself, in its journal's words, 'to the fundamental elements of social life—work and material culture, class relations and politics, sex divisions and marriage, family, school, and home'.

Lastly, there has been a great growth in local history. At first this was especially rural, where the method had deeper roots. Gough has been mentioned as one type; folklore collecting provided another; and a remarkable example was also set by the Women's Institute histories from the 1920s. These were village surveys, based partly on the example of the Scottish Statistical Accounts, but equally—through the influence of C. V. Butler— on Rowntree's social surveys. Joan Wake's *How to compile a history and present day record of village life* (1925) was written for the Women's Institute surveys, and gives excellent advice both on documentary research and the use of interviews to collect information from old people: on farming methods, tenancies, wages, trades and industry, transport, emigrants, schools, clubs, friendly societies, trade unions, health, food, religion, and crime; old stories, folklore, songs, and games; and personal reminiscences. 'Why not have "reminiscence parties"—when each in turn would recall and relate his or her experiences, while someone took them down in shorthand?' she suggested. After the Second World War Women's Institutes or Old People's Welfare Councils in many counties sponsored essay competitions; extracts have been published as Pat Barr's *I Remember* (1970) as well as in local booklets. It is partly from this strong tradition of local history, as well as from his understanding of folklore and of work experience, that the work of George Ewart Evans springs, especially in his first book, *Ask the Fellows who Cut the Hay* (1956). In its title and introduction this village study in fact constitutes the first appeal for the present English oral history movement. In more recent years, however, there has been a still stronger flowering of local community-history projects in the inner cities, some self-supporting, others funded through urban aid or, with the mounting unemployment of the 1980s, through the Manpower Services Commission.

Oral history, in short, has grown where there was a surviving tradition of field-work within history itself, as with political history, labour, and local history, or where historians have been brought into contact with other field-work disciplines such as sociology, anthropology, or dialect and folklore research. Its geographical distribution also reflects the availability of money for field-work: hence the high concentration in North America and north-west Europe. For the same reason government sponsorhip, especially of folklore collection, but also through unemployment schemes, radio archives, and social science research councils, have been key influences in most countries. In the United States some major government projects exist, but they chiefly concern the military forces and the experience of war. As a result, private funding has been dominant, with an emphasis on the recording of just those people who are most likely to leave written records, the national and local élites. There are even oral history projects on the fund-giving foundations themselves. Thus the patterns of sponsorship—and, it could be argued, the political assumptions which lie behind them—have also been key factors in shaping different national developments.

There is, however, one more factor: the nature of opposition. The system of private funding in America has had, in this respect, the happy consequence of allowing oral historians to go their own way, loosely attached to local universities, colleges, and libraries; although less fortunately it has led to the typical American oral historian being primarily an archivist and collector rather than a historian as such. In Britain, by contrast, a sharper struggle for resources and recognition was inevitable. With the economic recession and public spending cuts from the mid-1970s, any new claimant for scarce public funds was bound to meet opposition. Even the Social Science Research Council from cautious support had by 1976 switched to an openly hostile policy of 'containment'.[43] And if in the research world this proved a very temporary setback, there can be no doubt that in the subsequent university cuts of the 1980s the newest developments proved most vulnerable, and the well-established political and economic history of the safely remote past best able to protect itself.

Where such opposition succeeds, the main damage will be to professional historians themselves. Oral history will be developed principally by sociologists, anthropologists, and folklorists within educational institutions, and by lay historians in the community. Professional historians will miss the stimulation of inter-disciplinary work, and of contact with their own basic constituency; and they will allow oral history to evolve in ways which disregard their own needs and standards. For example, the present inadequacy of archival facilities, and the consequent destruction of a high proportion of the oral evidence which is actually being collected, is likely to continue until the historical profession accepts that oral records are of as much value as written documentation. Thus while in advanced countries such as Canada, Australia, and the United States, federal and state archives have been collecting oral history material as part of their regular programmes since the 1950s, it was only in the 1980s that Britain set up its own National Sound Archive as part of the British Library and its oral history work has only just begun to develop.

Nevertheless, in the long run—and perhaps quite rapidly—the present hostility will probably dissolve, and professional historians will return to their earlier view of the acceptability of oral evidence as one of many kinds of historical source. The change in methods of communication which has ousted the paper document from its central role make this ultimately inevitable. And the opposition turns out on closer examination to be united by feeling rather than principle. Principles are cited, but they are contradictory, and derive from two extremes of the profession.

There are, first, chiefly in economic history and demography, those historians who wish to disregard *any* qualitative evidence which is not open to statistical analysis. As a school, they can be traced back to the 1920s when economic history was establishing its autonomy, and social history was moving from the impressionist elegance of G. M. Trevelyan towards the more severe standards of Georges Lefebvre with his slogan, *Il faut compter*. They were later reinforced by the neo-positivist hostility to traditional history of Popper, and then again by the exaggerated expectations of the 1960s from social science in general and the computer in particular. Very probably, as George Ewart Evans has suggested,

this reductivism has been so fashionable because of the unconscious wish to gain some of the kudos and respectability the natural sciences have won for themselves during this century; and it gains powerful support from the widespread assumption that modern science has the answer for everything that man really needs to know.[44]

But such high claims have themselves brought disillusionment. Statistical history can no more unravel the past unaided than sociology can provide answers to all current social problems. The best economic historians and demographers have of course always recognized this: like the *Annales* school in France, or in Britain K.H. Connell, who in his influential discussion of the post-famine demographic transformation of the Irish family used oral tradition collected by the Irish Folklore Commission as one key source of evidence. There has meanwhile been a reaction within sociology itself against a predominantly statistical methodology in survey analysis, and a return to life story interviewing in the field which has brought sociology closer to oral history. Thus the more extravagant hopes of the neo-positivist statistical school look increasingly dated. One can see more clearly how far Michael Anderson's analysis of *Family Structure in Nineteenth Century Lancashire* is distorted by sticking to a rigid economistic model of the family which allows, for a half-Catholic town in the decade of Chartist unrest, neither political, nor religious, nor psychological factors to be considered. And the daring acrobatics of an economic historian like R.W. Fogel, who will construct data when he cannot find it, and aspire to re-evaluate the entire experience of slavery with sets of tables, now seem sorties which reveal more of the pitfalls of the method than its strengths. It is difficult to believe that economic history and demography, which through their closeness to the social sciences are naturally more familiar with the interviewing method than most branches of history, and have indeed already produced some notable supporters of oral history, will remain long-term obstacles to its advance.[45]

The professional old guard looks at first sight more formidable. A. J. P. Taylor, for example, despite his awareness of the receding value of the written document, maintains his resistance to the interview method. 'In this matter I am an almost total sceptic . . . Old men drooling about their youth—No.'[46] And if the old-style documentary historian is likely to find history increasingly difficult to reconstruct from the twentieth century onwards, he has only to stay on the firmer earlier ground to which he is already well fastened. But the situation is in practice less fixed than it looks. The traditional historian, partly because he is suspicious of theories and prefers to construct his interpretation from individual pieces of evidence gathered wherever he can locate them, is at heart an eclectic. If he is suspicious of oral evidence, it is chiefly just because until very recently it was, to an extent which now seems difficult to recall, either hidden or unrecognized by him. Arthur Marwick in his *The Nature of History*, published in 1970, included a very catholic discussion of historical sources in his chapter on 'The Historian at Work', ranging from the accepted hierarchy of primary and secondary written sources to statistics, maps, buildings, landscape, imaginative literature, art, customs, and 'the *folkways* of the period'. He even argued that 'a history based exclusively on non-documentary sources, as say the history of an African community, may be a sketchier, less satisfactory history than one drawn from documents; but it is history all the same'. Yet he included no reference whatever to oral evidence as such. It seems unlikely that a similar passage today would not discuss both the interview method and oral tradition.[47] The awareness of these potential sources is now widespread, and awareness itself brings a degree of acceptance. In addition, oral history projects have created a number of archives, which are being used by research students and cited in their theses, frequently with the encouragement of their supervisors. For this new generation then, oral evidence is again counted among acceptable sources. And since it can be cited in their theses, they have become generally willing, when this seemed potentially worthwhile, to consider collecting such evidence themselves in direct field-work.

The fact is that the opposition to oral evidence is as much founded on feeling as on principle. The older generation of historians who hold the chairs and the purse-strings are instinctively apprehensive about the advent of a new method. It implies that they no longer command all the techniques of their profession. Hence the disparaging comments about young men tramping the streets with tape recorders, and the grasping of straws to justify their scepticism: usually a reminiscence (it should be noted) about the inaccuracy of either their own or some other person's memory. Beyond this there is—and not only among older scholars—a fear of the social experience of interviewing, of the need to come out of the closet and talk with ordinary people.[48] But time will temper most of these feelings: the old will be succeeded; and a widening number will themselves know the positive social and intellectual experience of oral history.

The discovery of 'oral history' by historians which is now under way is, then, unlikely to be obscured. And it is not only a discovery but a recovery. It gives history a future no longer tied to the cultural significance of the paper document. It also gives back to historians the oldest skill of their own craft.

# 3

# The Achievement of Oral History

HOW do we measure the achievement of oral history? Against a roll call of its long past: Herodotus, Bede, Clarendon, Scott, Michelet, Mayhew . . .? Or its present ambitions and diversity? It is not possible to mark any clear boundary around the work of a movement which brings together so many different kinds of specialists. The method of oral history is also used by many scholars, especially sociologists and anthropologists, who do not think of themselves as oral historians. The same is true of journalists. Yet all may be writing history; and they are certainly providing for it. And for different reasons professional historians are also unlikely to conceive of their work as 'oral history'. Quite properly, their focus is on a chosen historical problem rather than the methods used in solving it; and will normally choose to use oral evidence along with the other sources, rather than alone. The term 'oral history' is itself a contribution to this confusion:

. . . it implies a misleading analogy with already differentiated aspects of history—economic, agricultural, medical, legal, and so on. Whereas oral history can never be a 'compartment' of history in its own right, it is a technique that could conceivably be used in any branch of the discipline. The title also suggests—indeed invites—another hiving off when in fact it is clear to anyone who has taken oral evidence in the field over any length of time that compiling oral sources is an activity that points to the connectedness of all aspects of history and not to their divisions from each other.[1]

If the full potential of oral history is realized, it will result not so much in a specific list of titles to be found listed in a section of historical bibliographies, as in an underlying change in the way in which history is written and learnt, in its questions and its judgements, and in its texture. What follows is a discussion of simply one dimension of oral history—the impact of new oral evidence in existing fields of historical study—and the examples cited are deliberately limited to modern works. Even so, it is difficult to make any satisfactory balanced choice between, on the one hand, the considerable number of often brief articles, especially on research in progress, which are known through direct publication in journals and bibliographies of the formal oral history movement, and on the other, the infinite but often substantial publications in sociology, anthropology, folklore, contemporary history, politics, and biography which lie to its fringes. A full survey of each field in turn would indeed be impossibly lengthy, and this will be simply an illustrative discussion.

Let us begin with economic history. Few will need to match the boldness—in more than one sense—of historians of pre-colonial central Africa like Robert Harms, exploring the tributaries of the inner Congo in his own canoe, who have pieced together the emerging patterns of production, trade, and markets in their regions principally from communal and family oral traditions. The role of oral evidence in economic history has normally been relatively modest: first, as a corrective and supplement to existing sources, and secondly in opening up new problems for consideration. For some aspects of economic history, such as government policy, foreign trade, or banking and insurance, the existing documentation is abundant even if sometimes narrow in focus. But some of the major aggregate historical statistical indices, for example of real wages, of hours, and of productivity, are compilations resting to a quite considerable extent on either inadequate documentation or on absolute guesswork despite the confidence with which they are normally presented. They are the basis, for example, of the great debates on the standard of living in industrial Britain: but Elizabeth Roberts has demonstrated from interviews with working-class families in two Lancashire towns how many factors have been misconceived or completely left out of calculations for statistical indices of the standard of living. And the sources prove equally defective for studying the history of many major industries: take mining, for example. Christopher Storm-Clark has shown how existing documentary records are both insufficient and misleading. The mining industry before the late nineteenth century consisted chiefly of small, shallow, and often short-lived local pits; yet the evidence which survives is not merely scarce and fragmentary, but heavily biased towards the atypical large-scale capital-intensive pits and their associated settlements. The closure of pits and consequent destruction of their records from the inter-war depression years onwards, the unwillingness of owners to allow their examination, and the subsequently similar fears of the National Coal Board, have improved neither their availability nor their informative content. For his own research, Storm-Clark has therefore used interviewing partly to collect basic information about the technology and work organization of the type of pit whose records are missing. Interviews also supply much fuller evidence of the processes of recruitment into the pits and migration into mining districts than any colliery records. Perhaps most striking, however, has been their value in elucidating and correcting the very information which, at least for certain pits, Colliery Wage Books do supply on working hours and wages. Interviews indicate that for the individual miner hours remained very flexible; while the system of piecework payments divided between workgroups of miners was so complex and variable that the concept of a wage *rate* for the period before 1914 is 'almost entirely meaningless'.[2]

The same kinds of arguments for the value of oral evidence in relation to documents apply to other industries. Thus Allan Nevins's massive social and industrial biography of Henry Ford, his company, and the automobile industry, shows how oral evidence can bring out more clearly than documents the working methods of a great innovator. And for our own *Living the Fishing* (1983)—on an industry dominated by small firms and seasonal labour—interviewing proved the quickest way of constructing an outline local economic history of each community and each family enterprise, and also helped us to see some of the errors in the abundant government documentation and statistics, which had reflected local pride or evasion or guesswork in supplying information for the official records; but still more important, it gave us the vital information on the contrasts in entrepreneurial culture between communities, which helped to explain why some had died while others continued to thrive. Indeed, more generally, it is as important to understand, in contrast to the big success story, the small firm like a country town iron foundry which did *not* grow into a great company; and, a step further back, the rural craftsmen—wheelwrights, smiths, thatchers, and so on—for whom written documentation is still sparser, but for whom now exists abundant literature drawing considerably on oral sources. Again, it is often only oral evidence which allows adequate study of a transient economic activity which may be a vital part of the wider picture. Thus there are virtually no written records of itinerant trades—hawking, credit-drapery, market-trading, and so on—and even for the highly organized brewing industry, there was only the barest documentation of the regular organized seasonal migration of farm labourers from East Anglia to Burton-on-Trent.

The most sustained oral history work, of critical significance for economic history, has, however, concerned agriculture. Here again accounts, wage books, and diaries can normally only be found for the larger, and more technologically advanced farms. The very existence of such records denotes an unusual degree of efficiency. Even where records exist, the information provided on, for example, wage rates or work techniques is normally inadequate, and frequently either incomprehensible or misleading. To secure any reliable indication of the normal labour patterns or the variations in technological level within a particular district, oral evidence is essential. The collecting of such source material has been most systematically carried out in Wales and Scotland, but as sociology, anthropology, or folklore rather than as economic history. More recently there has been distinguished work by David Jenkins and by Eric Cregeen on the social economy of these regions. But the demonstration of the relevance of oral field-work to agricultural economic history was led by George Ewart Evans, in his studies of East Anglian agriculture, *The Horse in the Furrow*, *The Farm and the Village*, and especially *Where Beards Wag All*: its methods, from the large steam-powered farm to the smallholding; cattle and corn economies; dealers, farmers, and farm labourers.

Several of these studies hint at another form in which oral evidence is beginning to contribute to economic history: the study of the entrepreneur. Although there is abundant autobiographical material on the upper- and middle-class intelligentsia, such information on the manufacturing and business classes is extraordinarily sparse. Without it, questions such as the role of the family firm and the socialization and attitudes of entrepreneurs in British economic decline cannot be answered. But economic historians have conspicuously failed to follow the example of sociologists in collecting life histories from industrial managers and from petty entrepreneurs. Their studies have brought important new findings: the lack of ambition of English small businessmen in contrast to managers, for instance, and the absolutely crucial economic role played by their wives. But it remains a paradox that we know much more about why French bakers or Scottish fishermen drive themselves to the limits for such small economic returns than we do about major financiers, industrialists, and developers; and probably the most revealing single life story we have of a businessman is of an Italian-American fence dealing in stolen goods, recorded for a study of deviance. Clearly a major opportunity is being missed here.

There is also a potential link between economic history and the history of technological and scientific discovery, although at present those oral history studies which do exist in the history of science are more concerned with its socially prestigious forms. There are projects in the history of medicine and psychiatry. And David Edge has provided in his *Astronomy Transformed: The Emergence of Radio Astronomy in Britain* (1977), a penetrating analysis of the post-war growth of the most spectacular, expensive, and perhaps least socially relevant 'big science', radio astronomy. Partly through his own previous experience in the same science, he understood that the paucity of records left by scientists was no accident; they did not regard their own earlier gropings and mistakes as relevant to the history of science, which they believed proceeded in a rational sequence of discoveries. Through interview evidence he has been able to show that the true picture is very different: a story of dead-ends, of misunderstandings, and of discoveries by accident, within a social setting of acute rivalries, partly handled by group specialization, but sometimes leading to the deliberate concealment of information. This constitutes therefore an important contribution to the historical study of scientific method; in which the scientist himself, from cool, rational superman becomes a more human and more political animal.

The history of science is of course but one branch of intellectual history. A particularly interesting area is the history of religion, for oral sources can here be used to distinguish the beliefs and practices of ordinary adherents from those of their leaders. How far, for example, has religion shaped the values of the lower middle-class white-collar worker? It is possible also to examine the 'common religion', superstitions, and rituals at birth, or marriage and death of the non-religious—by their nature areas mostly out of the reach of recent institutional religious documentation. And since the relationship between economic development and the religious ideologies of entrepreneurs and their workforces has long been a key subject of historical debate, this provides another point where oral evidence can continue economic history. A re-evaluation of the arguments of Weber, Halévy, and E.P. Thompson on this issue is the focus of Robert Moore's *Pit-men, Preachers and Politics*. This study of a Durham mining valley shows the role which Primitive Methodism, with its emphasis on individual self-improvement, backed by the paternalism of local pit-owners, played in inhibiting the growth of militant class-consciousness among the miners, until its influence, along with the paternalism of the owners, collapsed in the face of the twentieth-century economic crisis of the industry. The account of religion, including the identification of those who were local adherents but not members of the chapels, depends heavily on oral evidence, and the combination of a painstaking local reconstruction with a general theoretical argument makes this book a significant landmark.

It also brings us to an area contingent on economic history, but especially significant for oral history in its own right—that of labour history. The range of work here has already been sufficient to justify a separate bibliographical essay. It runs from local booklets, and articles in journals such as the *Bulletin of the Society for the Study of Labour History* or *Radical America*, to substantial books, and archive collections on the scale of the

South Wales Miners' Library. The contribution of oral evidence can be seen in several different forms. The simplest is biographical. Even labour leaders do not normally leave substantial private records, so that oral evidence has proved of regular value in an undertaking such as John Saville and Joyce Bellamy's *Dictionary of Labour Biography*, as well as in individual studies. But it has also transformed the character of labour autobiography. Despite many exceptions, the typical labour autobiography was until quite recently written by a trade union secretary or parliamentarian about his public life, at best prefaced by a few brief pages on his childhood and first job. Through the combined influence of oral historians, especially in community history projects, and also of broadcasting, we now have life stories from a much wider range of authors: from local as well as national leaders, from the ordinary rank and file, and also from non-unionized workers; from women as well as from men; from labourers, domestic servants, sweated and casual workers, as well as from miners and labour aristocrats. Equally important, the content and language have shifted from the public life to the ordinary experience of work and family. A more intimate and anecdotal type of autobiography has emerged, leaving its mark on the published life story. Its influence can be clearly seen in the extracts from recent manuscript autobiographies included by John Burnett in his fine collection, *Useful Toil*. A very considerable number of similar oral autobiographies are now available in record offices and archives. A selection have so far been published, most often as small local booklets, but also as collections like Alice and Staughton Lynd's *Rank and File: Personal Histories by Working Class Organisers from America*, or the People's Autobiography of Hackney's *Working Lives, 1905–45*. There are also a growing number of remarkable printed autobiographies of the new kind, which started as oral recollections, like Arthur Randell's *Fenland Railwayman*, Angus Maclellan's *The Furrow Behind Me*, and Margaret Powell's *Below Stairs*. But still more powerful is Angela Hewins's *The Dillen*, an autobiographical masterpiece recorded direct from a man who could never have written it, but had a rare gift for the spoken word. An orphan brought up in a Stratford-on-Avon common lodging house among down-and-outs and prostitutes, he was apprenticed to a local builder through his great aunt's determination, but fell for an early marriage and failed to serve his full time. His life became a relentless struggle to feed his growing family as a casual labourer, turned to bitterness through his savage mutilation as a First World War soldier: an unknown life of labour, yet unforgettable, which could have come to us in no other way.

Oral evidence can also be used to amplify information on specific events in labour history, such as the evolution of an organization, or the course of a strike. An exceptional example to which we shall return is Peter Friedlander's study of *The Emergence of a UAW Local 1936–1939*—the unionization of a Detroit car factory—which he built up almost entirely from a very searching form of interview. More generally, the advantage gained has been both in the spread of informants and the broadening of information to cover more of ordinary experience; and to most labour historians it will matter more how it is used than whether it comes in writing or on a recording. This may be seen by comparing the rather directionless memories of industrial disputes in R. Leeson's *Strike: a live history, 1887–1971* with the more purposeful written reminiscences which conclude Jeffrey Skelley's *The General Strike*. But better still are the essays in the same volume by Peter Wyncoll and Hywel Francis, and Anthony Mason's book on *The General Strike in the North-East*, which construct an account from a combination of oral and written evidence. A series of strikes can be analysed in the same way, like the harvest strikes of Norfolk farmworkers; or the early unionization of women in woollen mills or car factories; or a sustained campaign like the Welsh miners' response to the Spanish Civil War. Another approach is that of the urgent salvage operation to rescue material, often immediately in the wake of a dispute like the Upper Clyde Shipbuilders' work-in, or A. Lane and K. Roberts' report on the *Strike at Pilkingtons*. Both approaches can be of different value, but there are two particular strengths in oral evidence of this kind. First, it can get beyond the formalities and heroics of contending leaderships, as represented in newspapers and records, to the more humdrum, confused reality and different standpoints within the rank and file, including that of the blacklegs. Sidney Pollard and Robert Turner's analysis of a profit-sharing Yorkshire woollen manufacturer's workforce and its attitudes handles a subject which would have otherwise proved impenetrable. And some of the most interesting work has been carried out on the workers who were *un*employed: both of their organizations, and their experience of life out of work—the long, fruitless search for a job, the pinching of food, the humiliation of welfare—an experience depressingly similar whether in North America, Australia, or Britain. The widest collections of such evidence are in Studs Terkel's *Hard Times* and Barry Broadfoot's *Ten Lost Years*; but a richer, more reflective analysis, showing the use of the life story at its best, is provided by Dennis Marsden and E. Duff's contemporary study, *Workless*.

A third form of oral labour history, which also runs parallel to sociological research, is the community study, to which we shall return later. The impact of oral history here can be suggested by contrasting the earlier sociological classic, Norman Dennis, F. Henriques, and C. Slaughter's *Coal is our Life* (1956), based on interviews, but largely dismissive of the historical material which they collected, with the more recently historical and sociological work of Robert Waller and Robert Moore, in which the retrospective reconstruction of class relationships and sense of community has been a major concern. The method has also allowed the extension of historical community study to much more sparsely documented occupations, such as the casual labourers, carters, quarrymen, and laundrywomen of Raphael Samuel's 'Quarry Roughs'.

Finally, oral evidence has a special value to the labour historian concerned with the work process itself—not merely its technology, which we have discussed earlier, but the experience of work and the social relationships which follow from it. The experience of work is the concern of Studs Terkel's classic masterpiece *Working*. As with all his books, the effect is made not by explicit argument, but from cumulative interview extracts. It is a thick book: 600 pages in which 130 Americans pour out their work stories; old and young; real-estate woman, priest, factory owner, industrial spy, airline stewardess, hair stylist, bar pianist, strip-miner, car-welder, truck-driver, policeman, garbage man, washroom attendant . . . I know no other book which conveys so vividly the feeling of so many different kinds of job: the incessant, relentless tensions of the telephone receptionist; the loneliness of a top consultant struggling to survive in the jungle of management; the steel-millworker who would like the names of the workmen to be inscribed on what they make ('Somebody built the pyramids . . .') and short of this leaves here and there 'a little dent . . . a mistake, mine . . . my signature on 'em, too'.[3] One constructs one's own interpretations, although Studs Terkel no doubt has a shrewd idea of how they are likely to shape.

Much more clearly articulated studies of this type of labour history have now been published: on Fiat car workers and Terni steelworkers in Italy, textile workers in Manchester, New England, French and Spanish telephone operators, Thames shipwrights and coopers, English and German domestic servants and miners—and much else. In Italy the search to understand working-class consciousness through the direct feelings of workers themselves has led on the one hand to outstandingly perceptive historical studies such as Luisa Passerini's *Torino Operaia e Fascismo*, and on the other to the collection and publication of factory interviews, songs, and poetry—by journals like *I Giorni Cantati* and organizations such as the Istituto Ernesto di Martiro in Milan, and by autonomous workers' groups.

Two of the best studies once again concern mining. David Douglass's *Pit Life in County Durham*, based on a combination of documentary research and his own and others' experiences as miners, argues how the particular method of choosing workmates and workplaces in the Durham pits made for workers' control and rank and file militancy. George Ewart Evans sets out the system of the anthracite district of the South Wales coalfield, where the coal was near the surface, so that it was relatively easy for a small man to start his own drift mine; while its irregular geology gave special importance to the miner's skill. Owners and men lived and worked closely together. He then shows the impact of mechanization on the whole local social system, not merely destroying the status of the craftsmen, but also the close bond— sometimes paternal, sometimes exploitative—with the boys who formerly worked with them in their stalls, but now became a separate group beyond the control of the older generation. We have here an excellent example of how the exploration of a particular technical reorganization can illuminate its connections with other major processes of social change.

We have already touched, in considering the basis of changing class-relationships, upon a key aspect of political history; and the biography of labour leaders can be taken as another. But oral sources have a much more general relevance to political history. There is a strong case for their more extensive use in the historical study of the political attitudes of the unorganized, quiescent majority of the population. Neglect of this has meant that we still have only the sketchiest understanding of working-class Conservatism in Britain, despite its key role in political history. Similarly, oral evidence can provide much missing information on the attitudes of the rank and file of the parties: their reading, their social backgrounds and occupations, and so on.

That reconstructions of political organizations at the grass-roots level are possible, even where documentation is by definition largely non-existent, can be seen in William Hinton's uniquely illuminating unravelling in *Shenfan*, through the retrospective testimony of the villagers of Long Bow, of the complex feuds and devastating chaos of the Chinese cultural revolution. There are comparable studies of Mexico's long and confusing revolution, and also of European resistance to Fascism and underground political movements during the Second World War. The outstanding examples are the local studies of the Partisans in north Italy, and the international research on Jewish resistance under the Nazi regimes, which now centres on Yad Washem in Jerusalem. These enterprises have however been responses to rare national disasters, which have transformed the whole meaning of political history. The story of the concentration camps, whether told by survivors, or collaborators, or the children of victims, proves still as harrowing for those who tell as those who hear. But there are parallels in the memories of Hiroshima survivors of atomic bombing. And only a little less painful is the history of Spain under Fascism, the fear which drove men to live the best part of their lives in hiding, and the manifold, confusing, ambivalent experience of civil war for ordinary townspeople and peasants, men and women, winners and losers, which has been so brilliantly conveyed through oral history in Ronald Fraser's *Blood of Spain*.

Oral sources have more commonly been used for two much more limited purposes. First, there are studies of very recent political events which cannot possibly be satisfactorily analysed through written records. This has been a typical mode of American oral history, as, for example, William Manchester's *The Death of a President*, which drew on over 250 interviews, or W. H. Van Voris's *Violence in Ulster: an Oral Documentary*. Even where such works are simply high-quality journalism, they provided vital material for future historians. Secondly, there is biography. Here again the most striking instances are from the Americas, such as T. Harry Williams's *Huey Long,* or Valentina da Rocha Lima's collage portrait of the Brazilian political leader *Getulio*, each built from over 200 interviews. But the method, if less publicized, is also normally used by British political biographers: characteristically, in an informal and exploratory fashion. Martin Gilbert's volumes on Churchill provide an excellent recent instance of the fruits of this approach. And on occasion, a British political biographer finds the need to go further in the use of oral sources. Bernard Donoughue and George Jones interviewed over 300 people for their *Herbert Morrison: Portrait of a Politician*. 'We were forced from the beginning to resort to interviews because of the lack of certain other documentary sources. Morrison himself left very few papers, having burned the majority of them when moving house late in his life. The official papers for the 1945–51 government, in which he played a dominant role, are also not available because of the thirty-year rule.' Turning to interviews 'in some desperation . . . we were rapidly converted to appreciating their enormous value. They proved to be not just a stop-gap substitute for better sources, but a quite distinctly valuable source in themselves.' In particular, it proved possible to build up a much fuller range 'of perspectives and insights to the man . . . his virtues and his vices, and the extent to which the one was so often the reverse side of the coin to the other'. An early political life, so often skipped over by a biographer, could be reconstructed in remarkable detail. And throughout his career, Morrison could be revealed at work, as a Minister or in local government, through 'the various groups of people on whom he made an impact: his political associates, his political opponents, the civil servants working with him, the people at the grass roots who were supporting him or on the receiving end of his policies'.[4] The result, it can be added, is a biography which is not merely unusually rounded in itself, but has also created significant new historical source material for the future.

The concerns of political history extend beyond domestic events and biographies—very obviously in the case of Britain, which in the early twentieth century was an imperial power controlling a quarter of the world's surface and a colonial population of some 400 million people. There are several large-scale American and British projects collecting oral evidence in the field of military history. Once again, they are particularly important in illuminating ordinary experience, like 'life on the lower deck' of the navy, in the barrack-room, or of the black soldier on the Second World War battlefields. Similarly, they can provide otherwise unobtainable information on anti-war activities by conscientious objectors, passive resistance, sabotage, or outright mutiny within the forces. Parallel enterprises in colonial history concern themselves with the civil administration. The Cambridge South Asian Archive has focused on India, the Oxford Colonial Records Project on Africa. The fascination of this type of colonial social history has become widely known through Michael Mason's radio programmes on the British in India, and their printed sequel, Charles Allen's *Plain Tales from the Raj*. Through them, as in no other way, one may enter the strange, caste- and class-ridden world of the imperial white élite: the messes and homes of the officers and soldiers of the Indian Army, the pilots of the Calcutta river, the 'heavenborn' of the Indian Civil Service, their brothels, mistresses, and 'Memsahib the Wives and Daughters of the Raj'.

This is but one side of the story. The other concerns the people who were themselves colonized. The rarity of individual voices from among them still remains striking. The European anthropologists who followed the colonizers had none of the American interest in life stories, and oral autobiographies such as James Freeman's *Untouchable*, Indian outcaste, labourer and pimp, or *Kiki*, child in a New Guinea village of cannibals, or Mary Smith's *Baba of Karo*, the personal story of a Muslim Hausa woman in purdah and her marriages, divorces, and co-wives, remain as unusual as they are outstanding. Yet in Africa especially, above all for political history, but potentially equally for social history, oral sources play a crucial role. Documentation, although certainly present, is much less prolific than that of societies which became literate earlier, while oral source material is abundant. It has been systematically used by historians of Africa since the 1950s, with an increasingly sophisticated methodology, including the development of special techniques for the establishing of chronologies of oral traditions which quite often reach back to the sixteenth century, and in some cases still further. At first these traditions were understood essentially as orally transmitted documents, most valuable when they had survived intact from the remote past, so that the method required formal historical traditions and was more effectively used for the political history of relatively strongly organized African kingdoms, particularly in the period preceding their nineteenth-century colonization. Increasingly interest has shifted to the process by which oral traditions are varied and reassembled over time, and therefore to more diffused local political systems, where the very contradictions in the oral traditions of different communities or families provide the clues from which past political struggles and migration movements can be worked out. David Cohen's *Womunafu's Bonafu* and John Lamphear's *The Traditional History of the Jie* are remarkable histories of small forest and hill peoples in Uganda reconstructed in this way, while Paul Irwin's *Liptako Speaks* equally deftly exploits the contradictions in what he learnt among a savannah people on the upper Niger. The symbolic and social interpretation of origin myths has brought new meanings from them too, not only from an anthropologist like Steven Feierman in his *The Shambaa Kingdom*, but also from historians like Roy Willis, who in his *A State in the Making* pins the Fipa myth to the moment when these mountain

Tanzanians shifted from slash and burn to compost agriculture. Ironically the sheer ingenuity required to establish the elementary patterns of settlement and political power in pre-colonial Africa from oral sources seems to have diverted energies from exploiting their equal potential for the development of African social history, especially in combination with the direct life story evidence of the more recent past, which has already been used for the political history of nationalist movements.

For it is in social history, to which we now turn, that the relevance of oral evidence is most inescapable. My own *The Edwardians: the Remaking of British Society* was orginally conceived as an overall reassessment of the social history of the period, rather than a field-work venture. But I fairly soon discovered that although there was a wealth of printed publications from the early twentieth century, including numerous government papers, and some pioneering sociological studies, much of what I wished to know was either treated from a single, unsatisfactory perspective, or altogether ignored. Manuscript material could not fill these gaps because where normally accessible it simply enlarged on the bureaucratic perspectives already available in the printed sources. It was too recent a period for a satisfactory range of more personal documents to have reached the county record offices. I wanted to know what it was like to be a child or a parent at that time; how young people met and courted; how they lived together as husbands and wives; how they found jobs, moved between them; how they felt about work; how they saw their employers and fellow-workers; how they survived and felt when out of work; how class-consciousness varied between city, country, and occupations. None of these questions seemed answerable from conventional historical sources, but when Thea Vigne and I began to collect the evidence of, eventually, some 500 interviews, the richness of information available through this method was at once apparent. Indeed, much more was collected than could be exploited in a single book, so that in the end *The Edwardians* became as much a beginning as a conclusion, and the interviews collected for it have already provided the basis for other historical studies. Nevertheless it does indicate something of the overall scope of oral sources for social history.

Interviews provided a pervasive background to the book's interpretations; they were cited in all but two of twenty-two chapters; and some sections, particularly on the family, rely heavily on direct quotation. Equally important, as an antidote to the simplifications of an overall outline of social structure, I was able to present fourteen accounts of real Edwardian families, drawn from a range of classes and places over Britain, but obstinately individual—'the untidy reality upon which . . . both theoretical sociology and historical myth rest'.[5]

The field-work for *The Edwardians* was on a scale so far unusual, and in one respect for the moment unique: the choice of informants was guided by a 'quota sample', so that the men and women recorded broadly represent the regions, city and country, and occupational social classes of early twentieth-century Britain as a whole. Such a research plan is clearly not within the means of an individual scholar. The characteristic contribution of oral evidence has thus been not the essay in general social history, but the monograph, in various distinctive areas.

The first is rural social history. We have already seen how the way here was led by George Ewart Evans. His books are in their special way unsurpassable: direct yet subtle intertwinings of agricultural and economic history with cultural and community studies, portraits of individuals, and stories. In one section he may explore the contrasting social structure of an 'open' Suffolk village like Blaxhall of *Ask the Fellows Who Cut the Hay* with the paternalistic Helmingham of *Where Beards Wag All*. In another, with the eye of an anthropologist, he will suggest the significance of some superstition or tale concerning animals, or an odd dress custom like the 'breaching' of boys on leaving behind the long hair and petticoats of infancy. Perhaps best of all is his feeling for the life and the speech of the East Anglian farm labourer. Now and then he will point to the very particular quality of its syntax, its humour, its directness; and there is always the same care shown in his transcripts. In all these ways he sets an exacting standard for what has become one of the best-known areas for oral history. It is perhaps hardly surprising that when Ronald Blythe's *Akenfield: Portrait of an English Village* made an international literary success of Suffolk oral history, it was with less careful scholarship. Despite its title, *Akenfield* consists of life stories from several villages rather than a portrait of a single community; while in detail not only the language of the transcripts, but even its attachment to particular informants, cannot be trusted. The census of 'work in the village' also turns out to be an invention. But if as a model for sociology or history *Akenfield* has cut too many corners, it has proved indisputably successful in popularizing a new form of rural literature, a cross between the interview documentary and the novel. Nor can there be any doubt that oral evidence constitutes its real strength. Thus although the book opens with an idyll of cottages around the parish church, the hard reality of a village labourer's life at once breaks through with the first section of recollections by the older farm workers. It also becomes possible to see the community from conflicting standpoints, both of generation and of class, as one hears in turn farm labourer and farmer, vicar and gravedigger, Tory magistrate and Labour agent. Above all, it succeeds through the immediacy with which the spoken word confronts a reader with the presence of the people themselves.

*Akenfield* has therefore, despite particular—and avoidable—defects, proved a stimulus to oral history for essentially the right reasons. More recently, through offering authentic voices from the Italian and the French peasantry, Nuto Revelli in *Il mondo dei vinti (The World of the Defeated)* and Pierre Jakez Hélias in *Le cheval d'orgueil (The Horse of Pride)* have likewise fired the imagination in their own countries. And *Akenfield* has been followed by other community studies, often pushing rural history well beyond the concerns which were possible when only documentary evidence was employed. Raphael Samuel's fine study of Headington Quarry concerns a squireless hamlet of migrant farm workers, diggers, builders, pedlars, poachers, and washerwomen which is largely undocumented just because it was so egalitarian and ill-controlled, but, he argues, an essential and far from uncommon element in the nineteenth-century rural social economy. Oral evidence also allows a much fuller treatment of women in rural history. Mary Chamberlain's *Fenwomen* is a village study, influenced by *Akenfield*, but drawn entirely from the evidence of women, and again revealing an often harsh reality in a community in which 'men were the masters': in family and school, courtship and childbirth, chapel and village society, in service, whether in the kitchens, or out weeding on the windswept black-earth fields. Here again, the use of oral sources brings at once a new dimension to history.

All these examples come from the southern and eastern countryside of England: the region of arable farming and hired labourers. The family farm regions of the north and west attracted scholars concerned for oral evidence much earlier: collectors of literature and folklore, especially in Wales, Scotland, and Ireland, but also sociologists and anthropologists. The result has been a series of outstanding community studies from C. M. Arensberg and S. T. Kimball's *Family and Community in Ireland* (1940) onwards, all based on oral field-work. Two of the most stimulating are W. M. Williams's successive books on *The Sociology of an English Village: Gosforth* in Cumbria and the Devon village of *Ashworthy*. In the first, his emphasis is on the recent erosion of a traditional, stable social system; but in the second he argues that rural society was always in flux, readjusting from external pressures, economic, technological, or political, as well as from the rise and fall of individuals and their families. James Littlejohn's *Westrigg* is also particularly relevant, because it provides a successful model for a community oral history as an alternative to *Akenfield*: a very effective analysis of the changes in local class-structure during the past sixty years, as farmers have bought their own holdings from the old landowning class, and the former dominance of the sheep-farming economy of the Scottish borders has given way before the advance of forestry. And in another study, Ian Carter seeks to explain why farm workers in north-east Scotland, in contrast to their English equivalents, were not deferential in their social attitudes—yet failed to unionize. Social historians of these regions are now using oral sources too. In the Scottish Highlands, for a social history of the island of Tiree, Eric Cregeen has used oral sources not only as his major evidence for the beliefs and customs of the people, and for their accounts of the conflicts between the landowner and his factor and the community of crofters, and to balance the documents of an agriculturally 'improving' landlord with the continuing working system on the land; but more astonishingly to build up a picture of personalities, family relationships, occupations, and migrations from the *mid*-nineteenth century, with the result that the bare listings of the 1851 census are not merely enriched and interconnected, but given a time dimension, so overcoming one of their most serious limitations as historical evidence. David Jenkins was also able to draw on some remarkably detailed oral information for his *The Agricultural Community in South West Wales at the turn of the Twentieth Century*, allowing him to construct a meticulous and stimulating account of a local system of the division of labour and status which included, besides the wage-relations normal between farmers and servants, a type of 'work-debt' or labour service at the corn harvest taken in return for the provision of cottage potato-land.

The potential impact of oral evidence is equally strong if we turn from rural to urban history. Here, however, at first it generally produced new source material, rather than new forms of analysis. An exception was Richard Hoggart's classic study of the impact of magazines, films, and the other mass media on the culture and moral relationships of the working-class city community. *The Uses of Literacy* draws heavily on Hoggart's own recollections over forty years of a childhood in northern England. It is more explicitly oral history—one chapter is headed 'An Oral Tradition: Resistance and Adaptation: A Formal Way of Life'—in its attempt to examine working-class speech conventions in relation to social change. Hoggart's influence here, however, through his emphasis on the limitations of working-class speech, proved as much a handicap as a help, and still has its ramifications in oral history. It has provided an explanatory theme for Jeremy Seabrook's depressing studies of the prejudice and narrowness of the urban working classes, *The Unprivileged* and *City Close-Up*. Both of these are partly historical, the first an autobiographical family view from Northampton, the more recent from the northern mill-town of Blackburn; and if a useful counter to cosy romanticism, they seem too much shaped by bitter comment and tendentious interviewing by the author. Hoggart's theme was also taken up by educationists, and the doctrine that working-class speech was a crushing impediment to understanding was developed by Basil Bernstein from this basis. But more typical was the local community history, such as the pioneering booklets of Centreprise in East London and the linked collecting of oral memories and family photographs for neighbourhood histories like *Trafford Park* by Manchester Studies. Two notably good instances are North American: *Steveston Recollected*, an account of a Japanese-Canadian fishing town on the fringe of Vancouver, published by the British Columbian Provincial Archives, and the Boston Bicentenary Neighbourhood History Series, produced by a series of fifteen neighbourhood committees across the city, which combined library research with locating photographs and interviewing residents. Both illustrate the characteristic value of oral evidence in providing an urban history source from another standpoint: for Steveston, that of an ethnic minority group in the city; for a Boston neighbourhood like *The South End*, a record of how a community viewed its own—eventually successful—struggle to win a reprieve from wholesale slum clearance and redevelopment. There are vivid images of the detail of urban life—lodging houses, bars, prize fighters, and drunks—and the key connecting threads of migration and work.

The richness of this evidence, both for British and American cities, would now be disputed by few urban historians. Indeed, the great two-volume *summa*, Michael Wolff and H. J. Dyos's *The Victorian City*, included a chapter of portraits from four British cities drawn from my own interviews for *The Edwardians*, significantly captioned 'Voices from Within'. It is used particularly effectively in 'The Making of Modern London' series by Steve Humphries as illustration and sometimes as a counter to the documentary evidence: thus Londoners turn out to have been much more scared by the Blitz than had been assumed. Nevertheless, the step from illustration to analysis has proved more difficult. This is partly because urban history has concentrated on the big cities, and here the community study makes least sense, because even when a neighbourhood can be identified with distinctive boundaries, its people will almost invariably look beyond it for work, services, and definitions of their place in the city's social structure. One solution, to take a single block or street and follow the movements of all its people inwards and outwards, has been followed by Jerry White with notable success in two books, *Rothschild Buildings* and *The Worst Street in North London*, one on an East London Jewish tenement court and the other on a street of casual labourers and petty thieves, deprived families, and common lodging houses. He has a sense of physical and social space rare among historians, which provides a firm foundation for each book; and through the framework of local economy, policing, welfare, and culture, he deftly weaves the individual and family lives of each of these tiny corners of the great city. The result is a microcosm of the metropolis: a compelling new model for urban history. An alternative, but more illustrative, approach is the portrait of a neighbourhood, of which the most convincing example is Studs Terkel's classic in the Chicago sociological tradition, *Division Street: America*. This was conceived around his own boyhood in Chicago's Near North Side, where his mother ran a rooming house for single men. But he found that his search for 'a cross-section of urban thought' could no longer be confined to a single neighbourhood, and it grew into a hunt across the entire city: 'with the scattering of the species, it had to be in the nature of guerrilla journalism'. His people talk about both their past and the present; family, ambitions, work, politics; and they are men and women of all ages: black and white homeowners and home-makers from the window-washer to the aristocrat; architects and ad-men, craftsmen, the hot-dog man, the men's mag girl; the Republican precinct captain cab-driver, bar landladies and the police; and the migrants—Appalachians, the Puerto Rican nightwatchman, the Greek pastry shop-owner, Jesus Lopez the steelman. *Division Street*, vibrant with the class, racial, and cultural variety of that struggling city, is undoubtedly one of the masterpieces of oral history.

The great cities have drawn the attention, if only because their social problems have been the most acute: but the majority of people continue to live in the smaller towns. Although much more manageable subjects for community studies, sociologists and oral historians have so far taken little interest in them. The most brilliant insights have come as chance by-products: a factory portrait such as *Amoskeag* also gives us an American company town, a single life like *The Dillen* the underworld of Stratford-on-Avon. But the pioneering series of sociological studies from the 1920s by the Lynds of America's *Middletown* and Margaret Stacey's much later *Tradition and Change: A Study of Banbury* have had few followers; and while local small town histories exist drawing on interview evidence, until very recently there were few of distinction. This alone makes Melvyn Bragg's oral history of Wigton in Cumbria, *Speak for England*, an important landmark. The social change in this part agricultural and part industrial town is set out through the voices of a cross-section of its people: miners and farmers, dog breeders and pigeon fanciers, councillors, schoolteachers, housewives, and shopkeepers. There are patchwork sections on particular periods: the Edwardian days dominated by the Big House on the hill with its peacocks; the young men who went to the First World War to fight under colonels who called them 'rubbish' and returned to the bewildered disillusionment and unemployment of the twenties; the beginning of better times for many ordinary people at the end of the thirties, and the subsequent post-Second World War move towards much greater comfort, security, and leisure. Another section focuses on Wigton's chief factory, from its first keen pioneers to a present in which the labour organizer has become personnel manager, and a disillusioned shop-floor worker can harangue the 'rat-race' of 'snakes', while a promoted apprentice finds his way round the problems of eating his first managerial lobster. There is also a set of eight much fuller individual lives. They include such characters as Dickie Lowther, semi-crippled ex-valet to the aristocracy, griffon-breeder, Scoutmaster, and ritualist. But in significant contrast to the city oral history, the tones of Wigton are generally less spectacular. The quiet push of working-class people towards improvement which they document is perhaps thus all the more significant for the urban historian.

Some of the most telling sections of *Speak for England* concern the social history of culture—religion, education, and leisure. This is another area in which the use of oral evidence has already made a considerable impact. We have referred earlier to work on the social history of religion. In the history of education the major contribution has so far been made by sociologists, such as

Brian Jackson and Dennis Marsden in their classic *Education and the Working Class*, based on life story interviews from their own town of Huddersfield. The linguistic oral studies related to education have also helped to fill the gap until recently left by dialect studies, which often collect source material of considerable historical value, but have concentrated on rural communities. More recently, especially in the United States, there has been a growing interest in urban language and oral modes. As a result, studies of urban folktales and folklore and even folk preaching have been added to the already numerous publications concerning rural superstitions, tales, and crafts. Of these rural studies, a particularly compelling and imaginative blend of social history, folklore, and anthropology is offered by George Ewart Evans and George Thomson's *The Leaping Hare*. An earlier classic, again very much on its own, is Iona and Peter Opie's *The Lore and Language of School Children*, which revealed an astonishing historical depth of oral tradition surviving in the contemporary school playground.

There is also a well-developed academic scholarship in the study of music and folk-song. Here, thanks especially to the work of Edward Ives in New England, we now have not only studies of traditional song and its general historical context, but also social and musical biographies of individual singers. At the same time, a powerful case has been made for the use of both folk-song and its urban music hall sequel for the understanding of working-class ideology as, for example, in attitudes to marriage, sex, or class. There has also been a shift from the once overwhelming preoccupation with the 'traditional' element in working-class culture. Here sociologists have again been influential through their studies of urban working-class leisure forms like, for example, the northern industrial bands portrayed by Brian Jackson and Dennis Marsden in *Working Class Community*. Since such leisure activities rarely leave many records, they cannot be seriously examined without oral evidence. There have been recent oral history studies of particular leisure forms such as jazz bands, kazoo bands, fairs, and baseball, and also of the role, more extensive in its social historical ramifications, of the public house.

Leisure, whether as a means of courting to the unmarried, or of escape from home to public house of the married man, leads towards family history. In this area of social history the impact of oral evidence is especially important, enabling the historian to consider critical questions which were previously closed. The most striking early examples were provided by the anthropologist Oscar Lewis, whose deeply moving portraits of Mexican families, such as *The Children of Sanchez* (1961), are rightly famous. But many of the most notable community studies have been as much concerned with family, as titles such as *Family and Community in Ireland, Family and Kinship in East London*, or *The Family and Social Change: a Study of Family and Kinship in a South Wales Town* demonstrate; and these, like the family sociology of Lee Rainwater's accounts of American working-class marriage in *And the Poor Get Children*, also depend on rich life story evidence.

It was the almost complete absence of earlier testimony of this kind from ordinary people which allowed leading family historians to propagate the notion that love between parents and children or between married couples was a novel, 'modern' development of the last two centuries. We have one rare glimpse of the intimacies of everyday family life in the Middle Ages, reconstructed by Le Roy Ladurie from the evidence of shepherd families in the Pyrenean hamlet of *Montaillou* (1975) who were being investigated for heresy: and who, after that, could still hold such superior assumptions? Similarly, in his powerfully argued attempt to disentangle the causes for the declining size of middle-class families in late nineteenth-century Britain, *Prosperity and Parenthood* (1954), J. A. Banks could only cite the opinions of medical specialists, novelists, and other writers, but for all their evidence remain with 'no idea' whether it could be taken as 'specifically representative of the actions and words' of wider social groups, or 'how most members of the middle classes . . . had begun to think'.[6] There was no archive oral evidence available for England such as K. H. Connell and Martine Segalen have used to study peasant sexuality and marriage in Ireland and France. But the recent work of Diana Gittins on 'Married Life and Birth Control between the Wars' has shown that it is possible through interviewing to discover why parents chose whether or not to have children, and how they learnt about the contraceptive means which they used; and she demonstrates that the diffusion theory, whereby family limitation was held to have spread to the working classes through middle-class influence, is seriously misleading.

Her findings first appeared in a special *Family History Issue* of *Oral History* which also included articles on child-rearing and on youth. Both these aspects of family history have been further developed since, as in John Gillis's account of courting and marriage ceremonies, and Steve Humphries's provocative explanation of working-class delinquency as a form of family self-help in *Hooligans or Rebels?* But the impact of the new evidence is perhaps most powerful of all in two books, by Elizabeth Roberts on three Lancashire towns and by Tamara Hareven on the other Manchester, New England, which both investigate the whole family cycle from childhood to old age. Elizabeth Roberts's discussion of the family care of the old decisively refutes earlier sociological suggestions that the exchange of help can be explained as a calculated response to self-interest. Tamara Hareven demolishes another more widespread sociological assumption, that the nuclear family corresponds to the needs of industrialized economies, by showing the continuing effectiveness of the extended family, both as an instrument for migration and the supply of labour over long distances, and as a buffer in crisis. But she builds on other theories to draw out the complex way in which the family and economy interact, and how the relationship between family structure, tensions between generations, and class-consciousness is continually reshaped by the moment in the cycle of economic boom and slump when each generation starts paid work: by the crossing of *Family Time and Industrial Time*. Her book is a conceptual landmark. And the mutual influence of family and economy is a theme raised by other historians investigating child-rearing, marriage, and also women's work. It leads us directly to women's history.

Here again, the potential of oral evidence is enormous, and its possibilities are only beginning to be explored. Women's history was until quite recently largely ignored by historians, partly because their lives have so often passed undocumented, tied to the home or to unorganized or temporary work. The more notable early contributions were thus typically biographical or documentary, like Sherna Gluck's *From Parlor to Prison* on the lives of American suffragettes, Mary Chamberlain's *Fenwomen*, or the looser collection in Sheila Rowbotham and Jean McCrindle's *Dutiful Daughters*. But there has also been important research in Britain on the leadership and the rank and file of the women's movement in the early twentieth century, as well as women in labour and socialist politics and the peace movement, especially by Jill Liddington; and on both sides of the Atlantic a whole series of studies on women at work—in fields, factories, domestic service, in wartime, on the frontier—and also, if less often, at home and in the family. Sheer neglect has given this whole field the excitement of a voyage of discovery. But as essays like those in *Our Work, Our Lives, Our Words* show, or Anna Bravo's writing on solidarity and loneliness among peasant women, this new history also challenges basic assumptions about social structure and inequality, the 'nature' of men and women, the roots of power between them, and the moulding of consciousness through both home and work. Much more will certainly follow.

The limitations of written documentation apply equally to other social groups at the margins of power. This is most obvious with deviant subcultures, which sociologists have long explored through collecting life stories. More recent American work has broadened to include drug addicts and sexual transvestites, while Tony Parker in Britain has moved similarly from the professional thief of *The Courage of His Convictions* to the incompetent institutionalized ex-soldier in *The Unknown Citizen* and the eight sex offenders of *The Twisting Lane*. And the historical insights which can be won through this approach are now vividly demonstrated through Raphael Samuel's extraordinary record of the slum childhood and violent criminal adulthood of Arthur Harding in *East End Underworld*.

Other minorities are the survivors of conquest, or traditional social outcasts. American Indians, Australian Aborigines, and gypsies in Europe are all persecuted minorities, misleadingly documented by a hostile majority, but preserving their own strong oral tradition, through which a more understanding approach to their past becomes possible. In a similar way, oral sources are being used to bring new dimensions to the history of Chinese and Japanese and other North American ethnic minority communities; and also to both European and American Jewish history.

There have been two forms especially in which the history of minority groups has been influenced by oral evidence. The first is the study of immigration. The example here was provided by the interview field-work of sociologists from the Chicago school onwards, but normally this has been to examine the problems of immigration as a form of social pathology. More recently both sociologists and historians using oral sources have moved towards a more balanced approach in historical work, examining the ordinary experience of immigration, the process of finding work, the assistance of kin and neighbours, the building of minority community institutions, and the continuance of previous cultural customs, as well as problems of racial tension and discrimination. It can also suggest—particularly by setting the direct evidence of personal experience against the generalized message of the community's own oral tradition—how distorted are some of the commonly held explanations of immigrant social patterns in terms of racial or cultural inheritance rather than the simple economics of class factors.

The second form is Black history: in Britain, perhaps still a branch of the first, but in the United States decidedly distinct. It offers a cluster of outstanding works with which to conclude our exploratory survey of the achievement of oral history. We may usefully at this point step back and ask, what is distinctive about them as history? What do they do, which could not otherwise have been done? Three things. First, they penetrate the otherwise inaccessible. Two come from the great city ghettoes of urban America. Paul Bullock's *Watts, the Aftermath* is an account of a mass confrontation in Los Angeles; while Alex Haley's *Autobiography of Malcolm X* has few equals for conveying the bitter richness of city life or as a powerful portrait of an individual leader. Nor did the illiterate rural Black communities leave records for future historians. William Montell's *The Saga of Coe Ridge* is the leading American example of a serious fully documented community study by its subject largely dependent on oral evidence: an account of a Black colony, settled on a remote hill spur after emancipation from slavery, surviving at first through subsistence farming and lumbering, but degenerating through lethal fights with neighbouring Whites over women, and driven as natural resources became exhausted into moonshining and bootlegging, so that eventually it was broken up by the county sheriff's revenue men. Secondly, where records do exist, oral evidence provides an essential corrective to them. This is especially true of the old rural South—where history matters, as nowhere else in America, because it is employed to justify or deny the claims of White supremacy. It was thus no mere accident that the rich interview material which had been collected in the 1920s and 1930s from former plantation slaves and their dependants remained unused by historians for more than three decades. This has now been remedied, not only by full publication of the slave narratives in eighteen volumes edited by George Rawick—thus constituting the most important collective autobiography yet published—but also by the admirable interpretative essay, *From Sundown to Sunup, The Making of the Black Community*, which constitutes an introductory volume. And similarly—to narrow the focus to a single case study to which we will need to return— Lawrence Goodwin was only able to discover, through oral evidence, the true story deliberately concealed by contemporary newspapers and records of how the White upper class used systematic violence to destroy the inter-racial populism of one Texas county in the 1890s.

Finally, oral evidence can achieve something more pervasive, and more fundamental to history. While historians study the actors of history from a distance, their characterizations of their lives, views, and actions will always risk being misdescriptions, projections of the historian's own experience and imagination: a scholarly form of fiction. Oral evidence, by transforming the 'objects' of study into 'subjects', makes for a history which is not just richer, more vivid and heart-rending, but *truer*. And this is why it is right to end with Theodore Rosengarten's *All God's*

*Dangers*, the autobiography of Nate Shaw, an illiterate Alabama sharecropper born in the 1880s, based on a 120 hours of recorded conversations: one of the most moving, and certainly the fullest, life story of an 'insignificant' person yet to come from oral history. By fruits such as these, one would gladly see the method judged.

# 4

# Evidence

How reliable is the evidence of oral history? The question will be familiar to any practising oral historian. Our first task here will be to take it at face value, and to see how oral evidence stands up when 'assessed and evaluated in exactly the same way that you evaluate any other kind of historical evidence'. But as we shall see, the question poses a false choice. If oral sources can indeed convey 'reliable' information, to treat them as 'simply one more document' is to ignore the special value which they have as subjective, spoken testimony.[1] We shall come back to that. But let us first take the question as it was intended.

We can begin by looking over the shoulder of 'The Historian at Work', as described by Arthur Marwick in his *The Nature of History* (1970). First he lists the 'accepted hierarchy' of sources: contemporary letters, informers' reports, depositions; parliamentary and press reports; social inquiries; diaries and autobiography —the last usually 'to be treated with an even greater circumspection' than the others. In considering these sources, the historian must first ensure that the document is authentic: that it is what it purports to be, rather than a subsequent forgery. Next follows the crucial problem: 'How did the document come into existence in the first place? Who exactly was the author, that is, apart from his name, what role in society did he play, what sort of person was he? What was his purpose in writing it? For example, an ambassador's report . . . may send home the kind of information he knows his home government wants to hear . . . Does a tax return give a fair account of real wealth, or will there not be a tendency on the side of the individual to conceal the extent of his possessions . . .?' Or in using 'an exciting on-the-spot account' from an author or newspaper reporter, 'how can we be certain that in fact he ever left his hotel bedroom? These, and many others, are the sort of questions historians must ask all the time of their primary sources: they are part of his basic expertise.'² We may note that the authors of documents, like historians, are assumed to be male. More important, many of the questions which have to be asked of the documents—whether they might be forgeries, who was their author, and for what social purpose were they produced—can be much more confidently answered for oral evidence, especially when it comes from a historian's own field-work, than for documents. But little indication is given of how any of these questions, either of identification or of bias, can be answered. It is only in the case of medieval forgery that a specific expertise is mentioned. Otherwise the historian's resources are the general rules in examining evidence: to look for internal consistency, to seek confirmation in other sources, and to be aware of potential bias.

These rules are in practice less observed than they should be. The oral historian has a considerable advantage here, in being able to draw on the experience of another discipline. Social investigators have long used interviews, so that there is an abundance of sociological discussion on the interview method, the sources of bias in it, and how these may be estimated and minimized. Discussion of the bias similarly inherent in all written documentation is by comparison sparse. These are few guides to be found to the faults in any of the modern historian's favourite quarries.

Newspapers present a characteristic example. Few historians would deny the bias in contemporary reporting or accept what the press presents at face value, but in using newspapers to reconstruct the past much less caution is normally shown. This is because they are rarely able to unravel the possible sources of distortion in old newspapers. We may know who the owner was, and perhaps identify his political or social bias, but whether the normally anonymous contributor of a particular piece shared that bias can scarcely ever be more than guessed. Thus the evidence which historians cite from newspapers suffers not only from the possibility of inaccuracy at its source, which is normally either an eyewitness account or an interview report by the journalist. It is also selected, shaped, and filtered through a particular, but to the historian uncertain, bias. For example, when Bonar Law made his famous speech to a vast Conservative rally

at Blenheim Palace in July 1912, declaring he would support
Ulster in resisting Irish Home Rule by force, there were slight
differences in the reports of the exact words which he chose to use
in its key phrases. These reporting differences may have been
accidental or intended. Not all modern historians use the version
which *The Times* printed next morning. Nor, however, is it the
custom to point out such variants, even in a book of 'documents',
or a biography of Bonar Law.[3] This instance tells us more about
the historian's normal practice than its consequences, because
the historical effect of Bonar Law's words was more through the
newspaper reports than through their direct impact at Blenheim.
But another example may show how newspaper evidence can be
systematically misleading as well as inaccurate. Lawrence Goodwin has used newspapers and other written sources in combination with interviews in a political study of a county of East Texas,
in which a Whites-only Democratic party ousted the inter-racial
Populists from power in the 1890s. It was impossible to tell from
the local Democratic press either how this happened or indeed
how the Populists had maintained support in the first place, and
who most of their political leaders had been. Goodwin was able
to discover three separate oral traditions from different political
standpoints in the community which, when linked with press
reports, showed that the Democratic countercoup had been
based on a systematic campaign of murder and intimidation. Not
only had the newspaper deliberately omitted the political significance of what it did report, but some of the 'events' reported had
not happened and were published as part of the intimidation.
One politician who was reported dead, for example, in fact
escaped his murderers and lived another thirty years.[4] But
Goodwin's refusal to rely on newspaper evidence is rare among
historians—and it has an interesting basis, as in an earlier career
he was a journalist himself.

Most historians would feel themselves closer to the heart of
things with correspondence. Certainly letters have the advantage
of often being the original communication itself. But this does
not free them from the problem of bias, or ensure that what
letters say is true, or conveys the real feelings of the writer. They
are in fact subject to the kinds of social influence which have been

observed in interviews, but in an exaggerated form, because a letter is rarely written to a recipient who is attempting to be neutral like an interviewer. Yet historians rarely stop to consider how far a particular letter has been shaped by the writer to meet the expectations of its envisaged recipient, whether a political enemy or a political friend, or a lover, or perhaps even the tax inspector. And if this is true of letters, it is much more so of such other primary sources as paid informers' reports, or depositions—the statements of evidence made in anticipation of a possible court hearing.

Printed autobiographies are another very commonly cited source. Here the problems of reliability are more generally acknowledged. Some are shared with the life-history oral interview. In A.J.P. Taylor's view, 'Written memoirs are a form of oral history set down to mislead historians' and are 'useless except for atmosphere'.[5] They lack some of the advantages of the interview, and offer little in compensation. The author cannot be cross-questioned, or asked to expand on points of special interest. The printed autobiography is a one-way communication, with its content definitely selected with the taste of the reading public in mind. It cannot be confidential. If it is intimate, it is with the consciousness of an audience, like an actor on the stage or in a film. As a public confession, it is controlled, and rarely includes anything which the author feels really discreditable. In those cases when it is possible to compare a confidential interview with a life story written for publication, there seems a consistent tendency to omit some of the most intimate detail, to forget the trouble with unruly children farther down the street, for example, which can be much more revealing than the rosy generalization that 'Children had more respect for their elders then'. Nevertheless, just because it is is printed rather than recorded on tape, many historians would feel happier citing a published autobiography than an interview.

Many of the classic sources for social historians, such as the census, registrations of birth, marriage, death, Royal Commissions, and social surveys like those of Booth and Rowntree, are themselves based on contemporary interviews. The authoritative volumes of Royal Commissions rest on a method which was shaky even when a Francis Place or a Beatrice Webb was not at work manipulating witnesses behind the scenes. They used a peculiarly intimidating form of interview, in which the lone informant was confronted by the whole committee—just like a widow seeking out-relief who faced the Board of Guardians.

Most basic social statistics are also derived from human exchanges and consequently rarely offer a simple record of mere facts. Émile Durkheim believed, when he wrote his classic study of *Suicide*, that it was possible to treat 'social facts as things': as immutable, absolute truth. But it is now accepted that the suicide statistics which he used vary as much with the degree to which suicide was regarded as a social disgrace to be covered up, as with the rate at which people killed themselves.[6] Similarly we know— from other, retrospective interviews—that the marriage registers of the late nineteenth and early twentieth century grossly under-estimate the marriage rates of those younger age groups who should have obtained parental consent to marry. Those who thought their parents might object simply misstated their ages to the registrars. The later figures show that the younger true rates were *double* those recorded at the time.[7] Food statistics, such as those of the consumption of different kinds of fish, were distorted by the need to market new types of fish under old names: it was common practice, for example, to sell cat-fish, weaver, task, or gurnett as haddock or filleted haddock. Figures for the proportion of the workforce which was skilled show startling discrepancies which are only explicable as social points of view: thus the census statistics, based on self-report, have remained high and slightly rising while those from employers' returns have plummeted. Similar problems affect even the recording of physical facts such as housing. The census definition of 'a room', used for measuring overcrowding, was a social one, which determined the exclusion of sculleries, and how substantial a partition was required before one room was counted as two. But social historians, perhaps because they have come to statistics relatively recently, much too easily fall into Durkheim's trap of treating them as 'things'.

This is true even of historical demographers. Here surely, one might hope to find historians dealing with hard facts. But take the table of 'Completed Family Size By Year of Marriage' from 1860 to 1960 confidently printed by E. A. Wrigley in his *Population and History*. This is based on various sets of retrospective interviews with mothers, and assumes their accuracy in remembering the number of live births which they had. But no allowance is made for the numbers of those children born who died in infancy or early childhood, so that the table does not measure the average number of children actually reared—the 'completed size' of family as experienced by its members. Because of high child mortality, the average size of family before 1900 was much smaller than the table suggests, and never actually as high as the so-called mean completed family size of the tabulation. In other words, 'completed family size' is a demographer's abstraction, not a social or historical fact. Statistically minded historians and sociologists have ignored this. They have displayed no awareness that while the trend in the table is beyond dispute, the actual figures—however critical for population studies—are not. They are *estimates*, which have been haphazardly subject to significant revisions in recent years by the Registrar-General, even for the years before 1914—so that mutually contradictory sets of tables can be found juxtaposed (without any explanation) in A. H. Halsey's recent statistical collection, *Trends in British Society Since 1900*.

Social statistics, in short, no more represent absolute facts than newspaper reports, private letters, or published biographies. Like recorded interview material, they all represent, either from individual standpoints or aggregated, the *social perception* of facts; and are all in addition subject to social pressures from the context in which they are obtained. With these forms of evidence, what we receive is *social meaning*, and it is this which must be evaluated.

Exactly the same caution ought to be felt by the historian faced, in some archive, by an array of packaged documents: deeds, agreements, accounts, labour books, letters, and so on. These documents and records certainly do *not* come to be available to the historian by accident. There was a social purpose behind both their original creation and their subsequent preservation. Historians who treat such finds as innocent deposits, like matter thrown up on a beach, simply invite self-deception. It is again necessary to consider how a piece of evidence was put together in the first place. Thus, for example, official information from School Board and County Council records does not suggest that women teachers were required to resign on marriage before the 1920s, when this became an official policy; but thereafter, records this as the consistent practice. Yet individual life stories document quite frequent requests to resign on marriage before 1914, as well as appointments of married women to posts during the operation of the bar. The official reports of the Poor Law Commissioners of the Labour Migration Scheme of the 1830s can similarly be shown, through alternative sources, to have grossly exaggerated the figures for the numbers of paupers removed, and to have quite falsely claimed that all of those removed had found work, in order to suggest that the scheme was succeeding. At another level, even such apparently accidental social documentaries as photographs and films are in fact quite carefully constructed. So much so that, for example, almost *all* the sound background to Second World War film is faked. And on rare occasions, one can discover how for the 'casual' family snapshot, everybody in the picture was forced to change out of their normal clothing. Not merely this, but a similar decision is made about what is *kept* for the album. And the same kind of weeding shapes the public archive. The process of discarding and confusing memories to fit modern needs, which has been identified as a form of genealogical conquest in African tradition, has its equivalent in the systematic, if half-conscious, doctoring of the record sets which is the practice of Western countries. One can only refute, as thoroughly misleading, Royden Harrison's assertion that written archive sources, 'the type of evidence upon which historians set the highest store', possess a special superiority over oral material, because they constitute 'a kind of primary evidence which takes the form of pieces of paper which have been bequeathed to us unintentionally, unselfconsciously; secreted by institutions or by persons in the course of their practical activities'. Contrary to his assertion, this *is* 'a matter of some superstitious prejudice in favour of the written over the spoken word'.[8]

The true distinctiveness of oral history evidence comes from quite different reasons. The first is that it *presents* itself in an oral form. As an immediate form of record this has drawbacks as well as advantages. It takes far longer to listen than to read and if the recording is to be cited in a book or article, it will need to be transcribed first. On the other hand, the recording is a far more reliable and accurate account of an encounter than a purely written record. All the exact words used are there as they were spoken; and added to them are social clues, the nuances of uncertainty, humour, or pretence, as well as the texture of dialect. It conveys all the distinctive qualities of oral rather than written communication—its human empathy or combativeness, its essentially tentative, unfinished nature. Because it will remain exactly the same afterwards, a text cannot be decisively refuted; that is why books are burnt. But a speaker can always be challenged immediately; and unlike writing, spoken testimony will never be repeated in exactly the same way. This very ambivalence brings it much closer to the human condition. Paradoxically, in one sense through freezing speech in a tape recording some of this quality is lost. Nevertheless, the tape is a far better and fuller record than can ever be found in the scribbled notes or filled-up schedule of the most honest interviewer, or still less in the official minutes of a meeting. We have seen earlier how the 'doctoring' of official records has become so accepted that even the Cabinet Minutes document less what happened in the Cabinet, than 'what the Civil Service wishes it to be believed happened'. This is equally true at the humblest level of the parish council. George Ewart Evans first became 'sceptical of official records' while he was himself a local councillor. 'Not that there was any blatant inaccuracy . . . But since the time of the meeting so recorded, a selective intelligence had been at work, omitting almost everything that did not contribute to fortifying the main decisions reached.' The result was a set of minutes 'streamlined to the point of appearing to be the record of a different meeting'.[9] In the same way, the notes of the interviewer seek to contribute to the survey's hypothesis, to fill in the blanks in the schedule. Or the record of an 'exchange of views' between politicians is purged of its damaging passages and slips. The uniquely telling accuracy of the recorded tape, as evidence, needs less arguing since Nixon tripped himself with it over Watergate.

In these instances, since the original communication was oral, the oral recording provides the most accurate document. Conversely, when the original was itself a written communication, as in a letter, that written letter must remain the best record. However, the distinction is commonly less clear, because we communicate through both means. Sometimes a 'sacred' moment defines a particular form as authoritative: the judge pronounces sentence, but the death warrant is signed; the priest says mass from the book, but the international agreement is signed as a treaty. But how do we classify a letter, orginally dictated to a secretary, checked through reading back, discovered by a historian in the recipient's private papers, and quoted aloud to students attending a history lecture? Or the private recollections of a person, widely read in recent history, recorded at an interview, transcribed, and returned with written comments? Or the particularly puzzling practices of the law courts, where proof is argued out through oral testimony and debate, and written documents are read aloud; yet, quite inconsistently, their own proceedings are never taped, but recorded in paraphrase by a clerk, and judges tend to take more account of written than oral evidence, as if the oral performance was a merely rhetorical drama justifying truths conveyed on paper?[10] Certainly in each case there are both oral and written links in the chain of transmission; and either can modify or corrupt the original. And in none is it obvious which the original is.

For some historical eras one can be more confident. Thus even after the Reformation in Europe the principal means of communication was oral. People in general perceived the world as much through the sound of fellow human beings, or animals, and also through smell, as with their eyes. For this era, the document is normally a subsidiary record. With the spread of literacy, and the increasing use of the letter, the newspaper, and the book, the dominant means of communication was through the written or printed word. The paper document can then be primary; word of mouth a subsidiary form. Today the printed word has again been displaced by a more powerful means of audio-visual communication, in television and film. The visual-verbal form has thus in turn become subsidiary; and as the telephone increasingly replaces the letter, the original in exchanges between individuals becomes once more the oral communication. There are of course, in each of these stages, differences between social classes, and between subjects of communication. But the main point is that the original of evidence is sometimes oral, and sometimes not, and equally may or may not present itself, after transmutations, in the same form; and neither oral nor written evidence can be said to be generally superior: it depends on the context.

The evidence of oral history is, however, also distinctive in being normally retrospective over a longer time span. This is not because its sources are spoken. On the contrary, the tape recorder makes it possible to take statements during or immediately after an event, while writing almost always demands more of an interval. And most written sources—whether from newspapers, court hearings, Royal Commission interviews, or committee minutes—are also retrospective. Neither contemporary nor historical evidence is a direct reflection of physical facts or behaviour. Facts and events are reported in a way which gives them social meaning. The information provided by interview evidence of relatively recent events, or current situations, can be assumed to lie somewhere between the actual social behaviour and the social expectations or norms of the time. But with interviews which go back further, there is the added possibility of distortions influenced by subsequent changes in values and norms, which may perhaps quite unconsciously alter perceptions. With time we would expect this danger to grow. In the same way, over time the reliance on memory apparently becomes more salient. To understand the extent of these problems we can fortunately turn to the literature of the social psychology of memory, and also of gerontology, for help.

It is generally accepted that the memory process depends on that of perception. In order to learn something, we have first to comprehend it. We learn it in categories, seeing how the information fits together, and this enables us to reconstruct it on a future occasion, or to reconstruct some approximation of what we comprehended. As Bartlett argued in his pioneering discussion

*Remembering* (1932), it is in fact only through this basic process of ordering that the human mind has overcome the tyranny of subjection to chronological memory. If we could not organize our perceptions, we would only be aware of what had most recently happened to us. Immediately after an event it does seem that we can remember a great deal more than later on. For a very short time we have something close to a photographic memory. But this only lasts for a matter of minutes. It is of crucial significance that this first phase is extremely brief. Then the selection process organizes the memory and establishes some kind of durable trace by a chemical process. Unfortunately the bio-chemical knowledge of the brain, despite its recent rapid advances, is not yet able to answer the particular questions which a social scientist would wish to ask about the memory process. However, a change takes place in the microstructure of the brain, which is certainly capable of resisting gross suppressions of mental activity like anaesthetics. Then, when the material is recovered, some kind of reverse process takes place: another situation is recognized, and the brain picks out the material and to a certain extent reconstructs it.

The process of discarding, which is the counterpart of selection, does continue over time. This clearly presents a problem for oral history. But the initial discarding is by far the most drastic and violent, and it affects any kind of contemporary witness. This can be demonstrated from the few studies over time that exist. Let us consider first an artificial laboratory experiment conducted by Dallenbach with pictures in 1913. Because it is artificial, like most laboratory tests it provides a poor index of the reliability of social memory. It is nevertheless striking that the number of errors remains more or less stable after the first few days. It suggests what may be a quite typical 'curve of forgetfulness'.

**Dallenbach's picture experiment, 1913[11]**

Fifteen students asked to scrutinize, and answer sixty questions on picture details:

| Days since saw picture | 0 | 5 | 15 | 45 |
|---|---|---|---|---|
| Number questions answered (av) | 59 | 57 | 57 | 57 |
| Number wrong answers (av) | 8 | 10 | 12 | 13 |

There are comparable findings from more recent Norwegian and American research on patterns of child-rearing, for which mothers have been re-interviewed over periods of up to six years. In each of these studies memory proved least reliable in recalling past attitudes, and best about practical matters such as feeding methods (95 per cent accurate after three years). Even after a few months, a mother's picture of childbirth and early infancy will differ a little from her original account. But when the time span is increased to six years, the inaccuracies show no significant increase. Similarly, for a later stage in life, it is now possible to juxtapose information given by the same person to oral historians over intervals as long as twenty years, and while differences of emphasis can be shown, the degree of consistency is the more notable feature.[12] Most significant, however, is the conclusion of a recent experiment devised to test, over a period of almost fifty years, the memories of 392 American high-school graduates for the names and faces of their contemporaries in classes of ninety or more students. They were first given eight minutes to list in unaided 'free recall' the names of all those who belonged to their class. They were then asked to pick, within eight seconds each time, first a series of names of their own classmates out of others; and then, similarly, pictures of their faces; and then, again with time limits, to match names to pictures and pictures to names. The findings are set out below.

**Classmates names and faces recalled**: percentage[13]

| Years since graduation | Free recall | Name recognition | Picture recognition | Name matching | Picture matching |
|---|---|---|---|---|---|
| 3 months | 52 | 91 | 90 | 89 | 94 |
| 9 months | 46 | 91 | 88 | 93 | 88 |
| 14 | 28 | 87 | 91 | 83 | 83 |
| 34 | 24 | 82 | 90 | 83 | 79 |
| 47 | 21 | 69 | 71 | 56 | 58 |

It is clear that on all counts the loss of memory during the first nine months is as great as that during the next thirty-four years.

Only beyond this do the tests suggest any sharp decline in average memory; and even this may be more due to declining speed in tests timed over seconds, and also to the effect on average performance of 'degenerative changes' among some of those in their seventies. Of equal importance is the finding that for those classmates who were considered friends, *no* decline in accuracy of recall can be traced, even over an interval of more than fifty years. The more significant a name or face, the more likely it is to be remembered; it is the others which a 'very slow forgetting process' gradually discards from the memory.

The memory process thus depends, not only upon individual comprehension, but also upon interest. Accurate memory is thus much more likely when it meets a social interest and need. It has been shown that illiterate Swazis, who might be thought to have particularly good memories because they can write nothing down, are no more capable of remembering messages for Europeans than Europeans are, but when they are asked about the exact descriptions and prices fetched by cattle sold a year previously, they can recite this, while the European who bought the cattle and noted their prices in his accounts cannot. Similarly, an 80-year-old Welshman in 1960 was asked for the names of the occupiers in 1900 of 108 agricultural holdings in his parish, and when his answers were checked against the parish electoral list, 106 proved correct. Reliability depends partly on whether the question interests an informant. It is lack of any intrinsic interest which vitiates many of the early laboratory experiments with memory—as well as some outside it. For example, Ian Hunter describes an experiment in which a meeting of the Cambridge Psychological Society was secretly tape-recorded. A fortnight later all the participants were asked whether they would set down what they remembered happening. On average they remembered under one in ten of the specific points made, and of those which they did recall, nearly half were incorrect. They incorporated statements from other meetings and occasions elsewhere. But the experiment demonstrates not so much the normal unreliability of memory, as the fact that this scholarly group, which relied on written material for its scientific progress, was meeting together chiefly for the social benefits of debate, interaction, and self-exhibition.[14]

Recalling is an active process. Bartlett wrote, perhaps with exaggeration: 'In a world of constantly changing environment, literal recall is extraordinarily unimportant. It is with remembering as it is with the stroke in a skilled game. Every time we make it, it has its own characteristics.'[15] He had in mind particularly how a story may be retold differently to various audiences in different situations, and how its recall can be stimulated by re-meeting an old acquaintance, or re-visiting the scene of some past event. A willingness to remember is also essential: a feature of memory which is especially relevant to interviewing. Conversely, recall can be prevented by unwillingness: either a conscious avoidance of distasteful facts or unconscious repression. It is of course a particular interest of psychology to revive these suppressed memories through the therapeutic interview.

Although laboratory experiments have succeeded in establishing the main elements of the memory process, they provide a poor guide to its reliability, because they take place in a social vacuum isolated from the needs and interests which normally stimulate remembering and recalling. One of Bartlett's classic experiments, for example, was to ask a group of ten Cambridge students to repeat to each other, in sequence, a Red Indian tale, 'The War of the Ghosts'. The final version retained no more than a few scraps of the original. But these students had no intrinsic interest in a story from another culture; for them it was just an experiment, whose outcome proved more interesting, as it happened, because of their own lack of accuracy. But there are epic tales among the illiterate peoples of Africa which have been passed down orally for at least 600 years. These tales are subject to variation when the social needs of their tellers and audiences have changed, but can be consistent enough for the original elements to be identified by studying the structure of different versions. And nearer at hand, Iona and Peter Opie have found very remarkable chains of transmission in their study of *The Lore and Language of School Children*. Because of the very rapid turnover of children in school, the links in the chain of transmission are much shorter than with adult oral traditions, so that a school jingle in 130 years will have passed down twenty generations of children, perhaps 300 tellers—equivalent to more than 500 years among adults. It is extraordinary, in view of this, how much survives. For example, in his *London Street Games in 1916*, Norman Douglas reported 137 child chants, and forty years later the Opies found 108 of them were still being chanted. And among the 'truce terms' used by children—whose accuracy is presumably especially important to them—are words like 'barley' and 'fains' which go back to the Middle Ages. They originated in adult vocabulary, but have been preserved only among children. 'Tiddly Winks the Barber' is a rhyme which children still repeat as it was originally composed, in 1878. The Opies have many nice examples of both survival and change.

On rare occasions one can show in an ordinary life story how a telling phrase has been retained. One of our own first interviews was with Bob Jaggard, an Essex farm worker born in 1882, who started work in 1894 on a farm, leading horses. Early on in the interview he said:

Men got 13 shillings a week and when I started work I went seven days a week for three shillings.

*Can you remember at that time whether you thought that was bad or good money?*

I knew it was bad money. Yes, they were put on.

*Did you feel there was anything you could do about this to get more money at that time?*

No, we didn't, that was just that. I can tell you right start, the old farmer what I worked for, he said a man carry a sack of wheat home every Saturday night was thirteen shillings . . .

Later on in Rider Haggard's *Rural England* I discovered that he had visited Bob Jaggard's village, Ardleigh, in 1901. A week's wages by this date had risen by eighteen pence. In Ardleigh he visited a Mr T. Smith, farming 240 acres, who had been there for fifty-one years. 'How could farmers get on', he asked Rider Haggard, 'when each man took the value of a sack of wheat; that is 14s. 6d. per week?'[16] Seventy years after Haggard's visit it was still possible to record the Ardleigh farmer's grumble, stuck in Bob Jaggard's mind.

One can of course match examples of recall with instances of misremembering; and individuals certainly differ in their ability to remember. But let us return to the normal memory process. How far is memory usually affected by increasing age? Young children, from birth up to the age of 4, have very little long-term memory at all. This is followed by a transitional stage up to the age of 11. Many children—more than half—retain a photographic type of memory, and also a great capacity for rote learning of a kind which is very unusual later in life, although it is retained by a small minority of adults. Some psychologists argue that the disappearance of photographic memory is connected with the onset of 'logical' thinking, although there are difficulties in demonstrating this. However, after the age of about 11, and especially after the age of 30, the immediate memory begins to show a progressive decline, so that it becomes increasingly difficult, for example, to retain a whole set of complex numerals in the head. On the other hand, the total memory store is increasing; it is as if one pushes out the other. Studies of vocabulary retention have shown that while for the most intelligent groups there is very little decline at all, for the average group tested a decline of memory sets in by the age of 30 and continues very slowly, but is never drastic before either terminal illness or senility is reached. Thus the problem of memory power is not much more serious for interviews with old people in normal health than it is with younger adults.

With this process of declining power in all adults the recent memory is first affected. Hunter writes: 'If there is, in the elderly person, an impairment of central nervous functioning, this favours recall of earlier as opposed to more recent events. With progressive impairment of a general neurological kind, recalling activities undergo progressive disorganization. That is, recall of recent events is impaired first.' There have indeed been statistical memory tests which, although in some respects methodologically doubtful, do suggest that if word associations are examined, nearly half go back to childhood or youth, and only a very tiny proportion are recent.[17]

The final stage in the development of memory commonly follows retirement, or some other traumatic process, such as widowhood. This is the phenomenon recognized by psychologists as 'life review': a sudden emergence of memories and of desire to remember, and a special candour which goes with a feeling that active life is over, achievement is completed. Thus in this final stage there is a major compensation for the longer interval and the selectivity of the memory process, in an increased willingness to remember, and commonly, too, a diminished concern with fitting the story to the social norms of the audience. Thus bias from both repression and distortion becomes a less inhibiting difficulty, for both teller and historian.

Interviewing the old, in short, raises no fundamental methodological issues which do not also apply to interviewing in general— and consequently to a whole range of familiar historical sources, as well as to those of the oral historian. It is to these issues which we must now turn. We shall explore approaches to interviewing more fully in a later chapter. Our concern here is with the degree of influence which the interview as a social relationship will inevitably have on the material which is collected through it.

The minimization of variance in answers due to differences of style between individual interviewers has long been the aim of social method. In the sociological handbooks this concern is often taken to self-defeating extremes. Ken Plummer, after charting all the possible errors listed, concludes that 'to purge research of all these "sources of bias" is to purge research of human life'. The real aim of the life-history sociologists or the oral historian should be to *reveal* sources of bias, rather than pretend they can be nullified, for instance, with 'a researcher without a face to give off feelings'.[18] But in precisely this sense, there is much we can learn from the experience of survey research. The key issue is how to introduce sufficient standardization without breaking the interview relationship through inhibiting self-expression. One approach has been to begin with a freer form of interviewing in order to explore the variety of responses obtainable, and then to follow up with a standardized survey, in which the exact words of questions and their sequence is predetermined. An alternative is to mix the two methods in each interview, encouraging the informant to free expression, but gradually introducing a standard set of questions in so far as these are not already covered. This protects the interview relationship, but makes the material less strictly comparable.

Since, in contrast to oral history, very little of this large-scale social survey interviewing is recorded, it is difficult to know how exactly interviewers normally follow such survey instructions. The rare tests which have been reported suggest that a third of the questions may be regularly altered in unacceptable ways. It is certainly clear that they carry into the interview both their own expectations and a social manner which affect their findings. For example, in one questionnaire survey interviewers asked women whether their husbands helped in purchasing house furnishings. The results differed depending on whether the interviewer's own husband helped or not. Those interviewers helped by their own husbands found this also the case for 60 per cent of their informants, while the other set, whose husbands did not, found only 45 per cent.[19] Some of the pressing and misinterpretation which went into these results would have been revealed in tape-recorded interviewing. There does not seem to be much doubt that much of the 'predictive' reliability of the contemporary social survey rests on the informal workings of both interviewers and analysts, who try to adjust their results towards what they themselves feel are credible conclusions. When nearly all of them are similarly mistaken in their expectations, as in the famous case of the 1948 Truman election victory, it is these informal workings rather than defects in the method itself which produce the wrong prediction.

Recording can help to expose and assess this kind of social bias. But the interviewer has a social presence, even when not revealing any explicit opinions which could influence the informant. There is a widely held image of an interviewer as a middle-class woman; and most informants have some idea of what her views are likely to be. This has some advantage, because the consequent bias in response can be more easily allowed for; and it can also be to some extent countered, by showing respect for the informant's own views. But there are interesting consequences when the image is unequivocably altered. For example, an American survey found that Black informants gave substantially different answers to some questions when asked by Black rather than White interviewers.[20]

**NORC survey 1942**

1,000 interviews with Black respondents, half interviewers Black, half White

| Question | Response | % when interviewer Black | % when interviewer White |
|---|---|---|---|
| Is enough being done in your neighbourhood to protect the people in case of air raid? | Yes | 21 | 40 |
| Who would a Negro go to, to get his rights? | To police | 2 | 15 |
|  | To law courts | 3 | 12 |
| What Negro newspaper do you usually read? | None | 35 | 51 |
| Who do you think should lead Negro troops? | Negro officers | 43 | 22 |

A parallel caution between races has been noted in Africa where, Vansina tells us, the White missionaries are expected to be interested in traditions. But they must not be told traditions that go against their teaching, because then they will criticize them, which will harm the prestige of the narrators, and will fight against them, which will harm the whole community.[21] In Europe an interviewer with a strong working-class accent, or a man rather than a woman, can expect to vary the social effect in—one hopes—a less drastic, but comparable manner.

It should be emphasized that it is not necessarily true that an interviewer of the same sex, class, or race will obtain more accurate information. If the social relationship in an interview becomes, or is from the start, a social bond, the danger towards social conformity in replies is increased. Nor does increased intimacy always bring less inhibition. It is remarkable, for instance, how many people, when stopped anonymously in the street by Mass Observation and asked questions about sex, were prepared to answer with a candour which is rare in the most intimate home interview.

The presence of others at an interview also has a marked effect. Boasting and exaggeration may be reduced, but the tendency to conform will be greatly increased. Howard Becker, when interviewing American medical students in groups, found that cynicism was the norm, but in private most students expressed idealistic feelings.[22] Sometimes a group meeting may be helpful, for example in bringing out conflicts in tradition about particular figures in a community's past from informants with different standpoints. And in a more personal interview, a husband or wife sometimes stimulates the other's memory, or corrects a mistake, or offers a different interpretation. An account of the division of domestic responsibilities given in such a situation, however, would usually be much less critical of the other's part. Equally it is noticeable that a group of old people will often emphasize a common view of the past, but if subsequently seen separately much more individual pictures may emerge.

Even when others are not present at the interview itself, their unseen presence outside may count. This is a particularly important influence in any tight-knit community. The insider and outsider have different difficulties here. The insider knows the way round, can be less easily fooled, understands the nuances, and starts with far more useful contacts and, hopefully, as an established person of good faith. All this has to be learnt and constructed by the outsider who, in the extreme case of a European student of African history, may not originally know the language, ethnography, or geography of the community. But there can be good in this too, for the outsider can ask for the obvious to be explained; while the insider, who may in fact be misinformed in assuming the answer, does not ask for fear of seeming foolish. The outsider also keeps an advantage in being outside the local social network, more easily maintaining a position of neutrality, and so may be spoken to in true confidentiality, with less subsequent anxiety. The insider's social situation is brought out by the experience of a student collecting information on her own Suffolk village. One retired horseman, explaining 'how hard things was in them days', told of how, when he had a child a few days old die, he carried the coffin to the churchyard himself in the early morning but had to pay a man to bury the child as he was due to start work at six. Due to the extra walk he reached work at six-thirty,

and in his pay at the end of the week the money had been taken off for the half-hour he was absent. The son of the farmer who deducted this money from his pay still farms in the same farm in the village today and a few days after that interview with the horseman he sent a message to me to say that he hoped he had not given me an unfavourable impression of the present farmer, as . . . 'times was hard then just after the (First World) War, there wasn't nothing else he could do really, can't expect him to pay a man for work he harn't done'.[23]

Andrew Roberts has emphasized the parallel handicaps of belonging to an African community:

Relations with the local people may well be more difficult than those of a white student. In so far as African students have kept up links with the land of their fathers, they come back to it as a full social personality, far more subject than a mere foreigner to the moral constraints of the society. If they ignore local custom in the cause of research, they (or their relations) will have to answer for the consequences. Through the web of kinship, they may well be caught up in conflicts which cause people to withhold information they might readily impart to a transitory white visitor. Besides, since independence, African students are rather more likely than whites to be suspected of being agents of central government.[24]

This is the extreme case of the problem, closer to the field-work situation of the anthropologist. One suspects that here in the long term the disadvantages of the European outsider may prove decisive. The social codes and layers of expressive meaning have to be penetrated as well as formal language itself. Even the very structure of conceptualization may be fundamentally different, and Western notions of time and space misleading. 'The scholar struggling to understand a foreign culture', Elizabeth Tonkin suggests, 'may eventually realize that what appear to be answers to the question "where did we come from?" are actually explaining "why we are here".'[25] The disadvantage of the insider in interpretation, on the other hand, is rather in the ease with which a community myth can be accepted at face value. Those others, often at the top and bottom end of the social scale, who carry a different viewpoint are not noticed. Nor can the social function of the myth be detected. For interpretation, as we shall see, this may prove much more revealing than the explicit messages which it conveys.

The message may also differ, depending on just where it is heard. Thus an interview at home will increase the pressure of 'respectable' home-centred ideals; an interview in a pub is more likely to emphasize dare-devilry and fun; and an interview in the workplace will introduce the influence of work conventions and attitudes. Linked with these changes in emphasis will be changes in language. A recording in a pub, for example, will often be festooned with swear-words; cross the home threshold, and the vocabulary will be transformed. Each might again vary if the interview was transformed from a confidential exchange to, at the other extreme, a television recording with technicians, glaring lamps, and a public audience beyond.

These then are some of the main influences from the interview situation. They are crucial: for they underlie the difficulties of any historian or sociologist in penetrating social reality, past or present. For the historian it is hardly possible to measure the extent of these difficulties, except when past errors come to light. But there are a number of sociological repeat surveys, which suggest how far any historical or contemporary evidence derived from interviews needs to be treated with care. In one study carried out by G. L. Palmer in Philadelphia in 1943, it was found that after only ten days 10 per cent of respondents reported their age differently by one year. Again, the Opinion Research Centre in 1949 made a comparison in Denver between interview survey material and local official records. It was found that once more 10 per cent of the answers were incompatible on age, 10–15 per cent on the possession of objects like library cards, cars, and the make of car, and 5 per cent even on the possession of a telephone. This study also calls into question the reliability of official statistics. A third example is, perhaps, more encouraging to the oral historian than to the organizer of the unrecorded questionnaire survey.

An experiment was carried out in New York during a survey of racial attitudes. Of the fifty respondents, eight were 'planted', and their interviews were secretly tape-recorded. Fifteen interviewers were employed, none of them full-time professionals. When the recorded interviews were analysed it was found that, out of the fifty questions supposed to be asked, on average each interviewer committed fourteen asking errors—that is changing or omitting the questions; thirteen probing errors; eight errors when recording the answers on the sheets; and then four simple 'cheats' (that is putting down an answer when none was given). One planted respondent acted as a 'hostile bigot', a type which could be expected to occur in most random samples. When faced by the bigot, half the interviewers invented half of what they put down on the questionnaire. If this is the kind of raw material which makes up the typical random sample questionnaire survey, we may well feel that the recorded interviews of an honest historian are likely, by comparison with much conventional evidence, to be unusually reliable.[26]

With this in mind, let us look at some cases in which the accuracy of retrospective material collected in large-scale surveys can be assessed. There is first the sociological study by P. M. Blau and O. D. Duncan of *The American Occupational Structure* (1967). The authors carried out a pre-test of 570 men in Chicago and tried to match their names against the census. They were only able to match 137, and in less than half of these did they find complete agreement of occupation and industry between the two sources. Tucking away as an appendix this rather damaging assessment of the foundations of their sophisticated statistical analysis, they argue that the discrepancies are partly due to high labour mobility in America (in 1945–6 this peaked at 12 per cent of all workers changing jobs), and partly—scarce comfort for historians—to the inaccuracies of a census at least as unreliable as their own survey. They cite a post-enumeration survey carried out by the Bureau of the Census to check its own results, which found that 17 per cent of the men were classified in a different major occupational group in the two surveys. This is a finding which might well be better known among statistically minded historians. Blau and Duncan were also able to show the differences between the census and their own survey were systematic. There was a tendency for labourers who appeared in the census to be described as craftsmen or technicians in their own questionnaire, but there was not a comparable error in the opposite direction. It was, on the other hand, reassuring to discover that the discrepancies, and so presumption of inaccuracy in retrospective interviewing, became *less* as the time interval increased. Men were more likely to describe correctly their father's occupation fifty-five than twenty-five years ago. For example, according to the census Blau and Duncan ought to have found that in 1940 12 per cent of the fathers were professional or managerial, and 17 per cent were farmers; but in fact far too many professionals and too few farmers were reported—20 per cent and 13 per cent respectively. But for 1910, when the census had recorded 11 per cent and 40 per cent, their own findings were much closer: 14 per cent and 38 per cent.[27] The reason for this retrospective increase in reliability is that an older man has fewer social reasons for wishing to mis-describe his father's occupation than a younger man. On some subjects it is therefore possible for the historian to get more reliable information from interviews than the contemporary sociological investigator.

A second large-scale retrospective survey is provided by David Butler and Donald Stokes's study of *Political Change in Britain* (1969). The historical information here is less closely analysed, but their tables of how each generation remembered their fathers' political views are clearly compatible with the broad picture which we have from other historical sources of a Labour Party rising rapidly at the end of the nineteenth century to oust the Liberals as the main contenders for power with the Conservatives. So are other figures showing that the Conservatives relied chiefly on the middle classes and the Church of England for their supporters, while their opponents depended upon Nonconformity and the working classes. Charles More's detailed statistical comparisons between oral history interviews and contemporary reports and census figures on skill and apprenticeship at the turn of the century are equally reassuring.[28] Such confirmations of established historical accounts clearly suggest that the retrospective survey in another field provides social information which in its broad divisions is reliable.

If we accept that memory is not so subject to error as to invalidate the usefulness of information gathered from retrospective interviewing, how do we overcome a rather different criticism—that our informants cannot be taken as typical or representative? Social surveys are normally based on carefully chosen samples, designed to secure as representative a group of informants as possible. They confront the oral historian with a dilemma. A survey whose informants are predetermined, and interviewed according to an inflexible schedule, will collect material of intrinsically lower quality. Some of the best potential informants will be missed, and others often less willing chosen in their place; while the interview itself cannot be sufficiently flexible to draw the most from them. On the other hand, one of the great advantages of oral history is that it enables the historian to counteract the bias in normal historical sources; the tendency, for example, for printed autobiography to come from the articulate professional or upper classes, or from labour leaders rather than the rank and file. Because of this, it is important to consider how far the oral historian could make use of some of the techniques of representative sampling developed by the sociologists.

The historian starts with a difficulty not shared by the sociologist. If the old people alive today were themselves a balanced cross-section of their generation in the past, in principle we should only need to draw a random sample from a list of their names. There would remain the practical difficulty of obtaining a fully reliable list which, unlike electoral registers, is rarely available. But we can be certain that such a 'random sample', even although providing the most certain form of present representativeness, would distort the past. It could take no account of migration, local or national, or of differential mortality. We know that people die much faster in some occupations than in others. Death rates can also be affected by personal losses, such as widowhood; by personal habits, such as smoking or drinking; perhaps by personality itself. Until a whole cohort of people has been studied from youth to age, we cannot be sure how far the cumulative effect of all these factors distorts the representativeness of the surviving group. But we do have measures of some of the most important differences between present and past, such as occupational and population distribution. This makes it possible for a large oral history project to rest on a frame which is, at least in some of its key dimensions, reliable.

For our own research project for *The Edwardians*, we recorded some 500 men and women, all born by 1906, and the earliest in 1872. Thea Vigne and I wanted to select a group representative as far as now possible of the Edwardian population as a whole, so

**Occupational and sex sample**

|  | Men | | Women | |
|---|---|---|---|---|
|  | *Occupied* | *Unoccupied* | *Occupied* | *Unoccupied* |
| Professional 18 | 4 | 3 | 4 | 7 |
| Employers and Managers 54 | 16 | 10 | 4 | 24 |
| Clerks and Foremen 28 | 10 | 2 | 2 | 14 |
| Skilled Manual 142 | 48 | 24 | 14 | 56 |
| Semi-skilled Manual 160 | 48 | 25 | 30 | 57 |
| Unskilled Manual 42 | 16 | 8 | 4 | 14 |
|  | 142 | 172 | 58 | 172 |
|  | 214 | | 230 | |

that we designed a 'quota sample'—a list of categories of various proportions into which people had to fit in order to be counted. The sample was based on the 1911 census and it totalled 444 persons. The proportion of men and women was as in 1911; so were the proportions who had then been living in the countryside, the towns, and the conurbations; and so too the balance between the main regions of England, Wales, and Scotland. We tried to ensure a proper class distribution by dividing the sample into six major occupational groups, taken from the adjusted census categories of Guy Routh's *Occupation and Pay, 1906–65* (1965). Those informants who were not working in 1911 went in as dependants of the chief breadwinner in the household, normally

**Geographical sample**

| | Rural 144 | Urban 200 | Conurbation 100 | |
|---|---|---|---|---|
| London | X | X | | |
| South East | | | X | |
| East Anglia | | | X | |
| West | | | X | |
| South West | | | X | |
| South Wales | | | X | |
| North Wales | | | X | |
| Midlands | | | | |
| Lancashire | | | | |
| Yorkshire | | | | |
| North West and North East | | | X | |
| Lowland Scotland | | | | |
| Highland Scotland | | | X | |

the father or husband. We had to carry out more interviews than the total 444 in order to fill the quotas, partly because some turned out to belong to a different classification than expected, and partly because not all were sufficiently complete.

Our aim was to present the people of Edwardian Britain who were alive in 1911, partly through those who survived, and partly through their children. And as a whole, the survey does succeed in this way, for the patterns which it produces by region and by class make sense. Some of the faults in the design of the quota itself could, on another occasion, be corrected. For example, we orginally failed to take account of the fact that because Edwardian women normally ceased working at marriage, the proportions of women working was far greater in some adult age groups than others.

It is all too possible to fill a category in the frame locally from a single social network which might, for example, exclude the less respectable. We therefore used a variety of means to find informants: personal contact, doctors' lists, welfare centres, visiting organizations, essay competitions, newspapers, and even chance encounter. We tried to notice the social bias which particular methods of contact could introduce, and counteract them. And there can be no doubt that the presence of the sample frame itself served to push the search for informants well beyond what would have otherwise seemed sufficient. The wholly unskilled, the 'rough' and 'unrespectable', for example, were again and again almost to the last moment socially invisible.

The quota sample carries one undeniable advantage over the random method. Since the choice of individual informants is not predetermined, there is no longer any need to force an interview on a respondent who remains unwilling, even after the purpose of the research and potential value of their contribution has been explained. Everything is to be gained from avoiding an interview which is likely to generate false material. But while it is clearly desirable to record only willing informants, there is another possible danger of going too far in the opposite direction, and recording only the exceptionally confident and articulate. Even within a particular social group or occupation, these may be a distinct stratum of leaders with their own culture and intellectual attitudes. Such informants are not merely unrepresentative, but can often prove less reliable. The more people are accustomed to presenting a professional public image, the less likely their personal recollections are to be candid; politicians are therefore particularly difficult witnesses. So are those who, through reading, have fixed upon a view of the past which they propagate professionally—such as historians and teachers. They can be the most insightful, but equally the most misleading sources. Indeed in African history Vansina suggests that the testimony of amateur collectors of oral tradition should always be avoided as 'quite worthless, because it is second-hand . . . "Listen to the words of the smith, do not listen to what the man who works the bellows has to say", as the Bushongo put it'. His ideal informant is a person still living the customary life, middle-aged or elderly, 'who recites traditions without too much hesitation, who understands their content but is not too brilliant—for if he were, one would suspect him of introducing distortions'.[29] The point—if not the patronizing tone—may be held relevant in Britain, too. If oral history is to be effectively representative, at all social levels, it is not just the unusually articulate and overtly reflective who must be recorded. Its essence is in conveying the words and feelings of ordinary people. The ideal choice is a broad one, but firmly grounded on the centre.

We may be certain in wishing to avoid interviews with unwilling informants. But what of those who are not so much unwilling as laconic, withdrawn? They will give the bones of a life story to a sympathetic interviewer, but never the most rewarding material. While they should clearly be included in any representative survey, what is lost if they are not deliberately sought out? This can be partly checked by observing whether their stories vary in any consistent direction from those of ordinary informants. We can secure a very rough indication from one American research project, which examined how far personality changed as people passed retirement and grew old—B. L. Neugarten's Kansas City Study of Adult Life. This categorized fifty-nine respondents before and after retirement as follows:[30]

## Kansas Study of Adult Life

Personalities of 59 respondents categorized

| Before retirement | Afterwards |
|---|---|
| (a)  19  well integrated | 16 socially active<br>3 socially disengaged, but calm, self-directed, contented |
| (b)  16  of 'armoured', 'defended' personality type, ambitious with high defences | 11 holding on—'I'll work until I'll drop'<br>5 constricted, closing themselves off from experience |
| (c)  13  of passive dependent type; rely on one or two people for emotional support | 8 satisfactory<br>5 apathetic, collapsed (widowed etc.) |
| (d)  11  of unintegrated type | 7 dissatisfied isolates<br>4 senile |

Although such findings must be treated with considerable caution, they do suggest that neither the withdrawn nor the unwilling informant is essential to secure a representative picture of earlier life experiences. The first, well-integrated group would present no difficulties. For the second and third groups, little would be lost by taking informants from only one of the two post-retirement sub-categories in each case. The incidence of widowhood, for example, is not consistently related to earlier life experience. And the last group would make unreliable informants in general. Of course this tentative study gives no measure of the similar changes in personality during childhood, youth, and earlier adulthood. If we are seeking evidence from childhood, we can assume with some confidence that there is *no* kind of family life which produced exclusively a single, uninterviewable type of personality, and is therefore inaccessible to the oral historian. Differences in personality type need to be taken into account, with, one hopes, perhaps eventually a more sophisticated measure of normal distributions, but they do not present an insuperable difficulty.

To meet the various problems raised by retrospective representativeness, the oral historian needs to develop, rather than the standardized random battery sample, a method of strategic sampling: a more tactical approach such as the 'theoretical sampling' advocated by Glaser and Anselm Strauss in their *The Discovery of Grounded Theory*. Various different approaches are worth considering. For many projects, as on an event, or about a small group of people, the issue is not representativeness, but who knows best. As the sociologist Herbert Blumer puts it, the search ought to be for *validity* rather than for reliability: 'a half dozen individuals with such knowledge constitute a far better "representative sample" than a thousand individuals who may be involved in the action that is being formed but who are not knowledgeable about that formation'. For other projects the whole objective might be to focus on a restricted group, for example interviewing members of the same family; or interviewing married couples and 'snowballing' by following up with their neighbours and friends. This would construct a picture of their social networks, attitudes, myths and memories, for which the very circularity of the enclosed group would be a strength rather than a weakness: as for instance in Yves Lequin's study of the collective memory of the metal-workers of Givors on the Rhône.[31] For a broader local study, the most appropriate method might be the 'community stratified sample', in which the aim is not to secure a mirror of its broad distributions, but to ensure the representation of all significant social layers within it. Or both aims might be met, by simultaneously working with two separate samples, the second devised on the 'quota sample' method which we used for our own national survey. No method of sampling can claim to be the best for all situations.

Concern for representativeness is essential if oral history is to realize its potential. The worst kind of oral history is that which begins and ends with the daily help. But it is equally important not to become obsessed with this issue, and lose sight of the substantive issues in developing methodology. And also of when they are best just forgotten. One of the deepest lessons of oral history is the uniqueness, as well as representativeness, of every life story. There are some so rare and vivid that they demand recording, whatever the plan. In a flash, we may be in another world, normally beyond even the most painstaking researcher: as in the experiences of a Glasgow girl, daughter of a proud artisan, a boilermaker who would sketch designs on the linoleum floor, but forced her and her brother—aged barely four and seven—out to sleep on the street to please their stepmother:

She just told us to get off, that's all. Yes, yes, shut the door on us. Gave us nothing . . . (for food), we used to steal the workmen's pieces to begin with. And then at other times, when we got too well known . . . we used to gather scrap and go to the rag store, get coppers and then go to these eating houses and get, maybe a bowl of soup . . . and then other times, Tommy and I used to go to a grocer's shop or a dairy, and I would ask the time while he was pinching scones off his table. That's how you lived. Between that and begging from door to door . . . (for clothes and shoes), we stole them in the cloakroom of the school . . .

I've slept under bridges, I've slept in people's doorways, with their carpets, and cats on the landing, and I've slept in watchmen's huts. I've slept in school shelters . . . In the dockyards, I've slept in the sheds, with the rats running round . . . I was nearly down the hold of a boat. In a truck of coal . . . I put the tarpolin over me, you see . . . They were just going to tip it into the hold of a boat—when the crane man saw me . . .

And then this old auntie . . . she used to chase my step-mother, because of what she was doing to me . . . She'd get a hold of us and take us away and wash our faces and soaped us all and do what she could for us. But she had to work. And always when she come back at night she couldna find us you know . . .

And this particular night, I'm asleep on her doorstep . . . But along came a lady, and I was sound asleep on this doorstep. She wakened me up, and asked a lot of questions . . . took me to her house, she carried me up, and she washed, cleaned me and put me in her own bed . . . She put me out the next morning, she had fed me, put all nice clothes on me . . . And she give me I think it was a penny . . . My father was in a close, in an entry, right facing . . . He whistled, and I looked . . . I thought that was great. And I goes to my daddy. You wouldna guess what happened. He struck me of everything I had . . . He went to the first pawn shop, pawned them. Pawned them. There I was left in the street again . . . I went back up to the woman. And I told her my daddy had taken all off me . . . Next thing I knew I was in court . . .

In the end—although not until the age of eight—she was taken into a church home, and her brother sent to a training ship.[32]

Just as in the selection of informants there are no absolute rules, but rather a number of factors to be taken into account, so in the end there are only useful guidelines to indicate when oral sources can be most reliably used, just as there are for other historical sources. The basic tests of reliability which we shall consider in detail in the final chapter—searching for internal consistency, cross-checking details from other sources, weighing evidence against a wider context—are just the same as for other sources. All are fallible and subject to bias, and each has varying strengths in different situations. In some contexts, oral evidence is the best; in others it is supplementary, or complementary, to that of other sources.

In the field of family history, for example, internal patterns of behaviour and relationships are generally inaccessible without oral evidence. The same is often true, in studying a strike, of the details of informal local organization, or of deviant behaviour such as blacklegging, or the normal devices like stealing fuel which helped families to survive with no income. The extreme case is the history of underground movement, such as the secret Jewish organizations in Nazi territory at the time of the Second World War. Yad Washem, the great Archive of the Holocaust in Jerusalem, has collected, besides some thirty million written documents relating to the persecution and exterminaton of Jewish communities in the Fascist period, more than 25,000 oral testimonies. Collection was begun as early as 1944, and, immediately after the end of the war, offices were set up in many parts of Germany and elsewhere for collecting evidence. Several of these centres are still active. They have collected a wide range of material on social and cultural life, partly in order to preserve some record of communities whose history would otherwise have died with them. Much more remarkable has been the ability to reconstruct, step by step, with the exactness and the patience which is needed for evidence which may need to be proved in court—and has regularly been tested in this way—accounts of both the persecution and resistance to it. When a large part of the Nuremberg trial evidence was subsequently lost by the Russians, Yad Washem was able to reconstruct three-quarters of the missing documents. As one of the archive's pioneers, Ball-Kaduri, knew from first-hand experience in Berlin, official documentation could not possibly provide an adequate record of the activity of Jewish leaders and their sympathizers who, in order to evade detection by the Gestapo, were forced always to meet in private, and to use spoken communication only. Yad Washem has indeed succeeded in preserving a history which, as he argued, written documents could never represent—'Was nicht in die archiven steht'.[33]

More often, the role of oral evidence is less dramatic, complementary or supplementary, reinterpreting and filling in gaps and weaknesses in the documents. The census of occupations, for example, is a very unsatisfactory record of secondary and part-time occupations. Through interviews it is possible to discover how a tradesman combined his craft with running a public house, or a casual worker took a series of occupations in a seasonal cycle, or many women described as housewives took in work at home or went out to part-time jobs. The labourer, 'that catch-all title favoured by the Census enumerators, turns out in many cases not to have been a labourer at all, but a man with a definite calling—a holder-up in the shipyards, a winch man at the docks, a well-digger or drainer in the countryside, a carrier or a freelance navvy'. Conversely, as a Divisional Inspector of Mines observed in 1912, 'You may have a perfectly good quarryman working three weeks or a month in a quarry, and another time he is a farm-labourer or working on some other work altogether.'[34] Such complexities could not be caught by the single entry of the census record, even if the enumerator was sensitive to them. And since for more recent periods the individual entries are anyway not available, in the meantime it may also be more accurate in a quantitative as well as a qualitative sense to use oral evidence:

Of what value would be the knowledge that 30 per cent of the workers in a particular plant were Polish, if we knew from previous investigations that this geographical unit was far too large to be meaningful? On the other hand, the response of an informant that a single department, say metal-finishing, possessed a work force that was 90 per cent Polish might be off by a few points, or even by as much as 10 or 15 per cent, but it would be far closer to the truth than the census estimate, which would be unable to go any further than specifying that 30 per cent of the workers in the plant were Polish.[35]

Similarly, while court records and newspapers might provide the best evidence for a dispute over common rights, or the numbers of poachers convicted month by month, oral sources could be essential to discover how the commons were normally used, or how the poaching system with its receivers, regulars, and casuals was actually organized. In his study of Headington Quarry, Raphael Samuel found oral history most useful in explaining the social structure and pattern of everyday life, least helpful in understanding a crisis, such as a political riot, and a prolonged dispute over school discipline, for which the contemporary documentation was richer. Interviews nevertheless probably do offer the best method for assessing the normal means used by teachers across the country for maintaining discipline in class. One critic of *The Edwardians*, contending that 'interesting reminiscences ought not to be offered as a substitute for a clear understanding', asserted that 'it is quite misleading to say that Edwardian teachers resorted *en masse* to corporal punishment. The debate over corporal punishment in State schools had begun in the 1890s, if not before, and many School Boards had begun to restrict its use even if the NUT protested at its complete abolition. A knowledge of the NUT's journal, *The Schoolmaster*, would have indicated this.' This journal does indeed show that there was debate. And one could also learn from the *School Board Chronicle* that teachers were demanding the right to use the cane. But it is certainly not possible to gain from these documents any kind of evidence of the extent to which corporal punishment was normally tolerated anywhere, as it is from the witness of the children themselves.[36]

As every experienced oral historian knows, however, the simple assertion and counterassertion that oral history sources are reliable or not, true or false for this or that purpose, obtained from this or that person, obscures the really interesting questions. The nature of memory brings many traps for the unwary, which often explains the cynicism of those less well informed about oral sources. Yet they also bring unexpected rewards to a historian who is prepared to appreciate the complexity with which reality and myth, 'objective' and 'subjective', are inextricably mixed in all human perception of the world, both individual and collective.

Remembering in an interview is a mutual process, which requires understanding on both sides. The historian always needs to sense how a question is being answered from *another* person's perspective. For example, general questioning on the good—or bad—old days will encourage subjective and collective myths and impressions; while detailed questions can draw out the particular facts and accounts of everyday life which the social historian may be seeking. But this does not mean that the myths and impressions lack any validity. The misunderstanding comes partly just because the historian is attempting to see change from another angle: the experience of one generation following another, rather than that of a single life cycle. When old people say that they enjoyed themselves more as children, or that neighbours were more friendly then, they may be perfectly properly evaluating their own lives, whether or not children find neighbours just as friendly today. Similarly historians too easily forget that most people are less interested in calendar years than themselves, and do not arrange their memories with dates as markers.

In general, one of the keys lies in mutual interest. Thus a man might be fascinated by the technological evolution of the motor engine through his years as a garage mechanic, but considerably less well informed about the upbringing of his children. But it is also true that over-interest can also present problems. An excessive concern with justifying the part they themselves played, as well as too much second-hand knowledge is, no doubt, one reason why politicians are apt, especially when not cross-questioned, to give somewhat casual accounts of major incidents. 'My experience is that memories are very fallible as a rule on specific events,' comments R. R. James; 'very illuminating on character and on atmosphere, matters on which documents are inadequate.'[37] But if personal pride and political interest make caution necessary in evaluating the recollections of politicians, with ordinary people sheer lack of interest is likely to affect their memories of national events. Melvyn Bragg, for example, discovered that it was pointless trying to collect information on important but historic events like Acts of Parliament or international incidents which to Wigton people were remote:

A man will talk of the Second World War, not in terms of Rommel or Montgomery or Eisenhower, but in a way in which everyone who served under those generals would understand. And poverty in the Thirties to a woman with six children would not be in terms of coalition governments and social legislation and trade union demands, but soup-kitchens, shoes for the family, the memory of a day's outing to the seaside—the common body of daily life.[38]

It is partly due to less interest, but also to a much briefer opportunity for incorporating them in the memory, that one can observe a *general* tendency for recurrent processes to be better remembered than single incidents. Thus a farm worker, in recalling an angry exchange with a farmer, may find the incident hard to place in time, and perhaps confuse the details with those of another similar occasion. But ask him about precisely how he handled his horses while ploughing, and it will be very, very rare for him to be wrong. A child's memories of King Edward VII's

Coronation Day are likely to be much less substantial than those of regular playmates and play places. In many events, people did not know at the time what was happening, so that their retrospective accounts will be as much based on what they learned from the news or from others as from their own participation. Indeed, just because such second-hand impressions may be more powerful than their own direct experience of the original fleeting incident, especially if they become a well-established part of community memory, some people come to believe that they actually saw an incident, such as an air raid, which they in fact experienced at second hand, through the newspapers or local talk. As David Jenkins observes of the village communities of south-west Wales, 'what is most accurately remembered is what has been periodically recalled and this is usually material concerning people; memory is much less reliable when it concerns events that neither recurred nor were consistently recalled'.[39]

It is certainly possible to reconstruct an event with oral evidence. But it is likely to prove a more difficult task, and unless this general tendency is understood, it may lead to serious misunderstandings. In his study of Henry Ford's development of the popular, mass-produced motor car, Allan Nevins was able to make rich use of oral evidence in giving body to the story which he found in the company's documents. Nevins comments, as a veteran oral historian, that 'any man's recollection of past events is untrustworthy'. But he knew how to use evidence effectively. For example, he could use it to establish Ford's own personal methods of working in the factory, like his avoidance of office work and letter-answering; and to separate the various roles in the team-work which went into the crucial Model T design. But in dating the introduction of the moving assembly line, he found that some Ford workers confused the first 'genuine attempt' of 1912 with an 'episodic . . . rope-hauling experiment' of four years earlier. Others correctly confirmed that there had not been a regular moving assembly line before the later date.[40]

This telescoping of two separate events into one in the memory is a very common phenomenon. For some purposes, the historian's task will be to try to separate them, delicately probing deeper; but for others, this very reorganization of the memory will be a precious indication of how a people's consciousness is constructed. Thus when Sandro Portelli interviewed Dante Bartolini, a veteran militant of the industrial town of Terni north of Rome, he told him how in 1943 the workers broke down the munition factory gates, seized all the weapons and escaped to the mountains to join the Partisans. Many of the workers had indeed joined the Partisans, where they established their own liberated zone, but they did not sack the factory in 1943; although Bartolini himself was one of those who seized arms in the factory after the arrest of the Italian Communist leader Togliatti in 1949. For Bartolini the resistance and the post-war industrial struggle are all part of a single history, eloquently conveyed in his symbolic story. In a similar spirit, almost half the steelworkers whom Portelli interviewed told the story of the post-war strikes with the killing of a worker by the police placed in 1953 rather than in 1949; and they also shifted its context from a peace demonstration to the three days of barricades and street fighting which followed the firing of 2,700 men from the steelworks. In fact nobody was killed in those three days in 1953. But as Portelli argues, the facts are not the interesting point about the episode. 'The death of Luigi Trastulli would not mean so much to the historian if it were remembered "right". After all, the death of a worker at the hands of the police in post-war Italy is not such an uncommon event . . . What makes it meaningful is the way it operates in people's memories.' Thirty, forty years on, in the 'longue durée' of memory, Trastulli's death still echoes in popular imagination. 'The facts that people remember (and forget) are themselves the stuff of which history is made.' The very subjectivity which some see as a weakness of oral sources can also make them uniquely valuable. For 'subjectivity is as much the business of history as the more visible "facts". What the informant believes is indeed a *fact* (that is, the *fact* that he or she believes it) just as much as what "really" happened.'[41]

We may illustrate this with a memory from British labour history which is again 'false', but nevertheless significant. Lindsay Morrison and Roy Hay were investigating a strike which took place in 1911 at the Singer factory in Glasgow. With the help of the only surviving worker from the workforce of the time (aged over 100) and his son:

we did piece together a story about how the Singer Company tried to break the strike. According to their version, which we subsequently heard from other independent sources, the company paid the Post Office to make a special delivery of postcards to all those on strike. The delivery was carried out on the Sunday evening and the postcards announced that all those who failed to report for work at starting time on the Monday morning would be considered to have left the employ of the company . . .

Now, we checked this story as far as we could from written sources, newspapers, a manuscript history of the company, and contemporary accounts. We found that there was a delivery of postcards but that they were made in the normal way and that the message they contained was somewhat different. The company said that when 60 per cent of the postcards they sent out had been returned, signifying the willingness of the workers to return on the previous terms, then the factory would be reopened. Obviously the pressure is here too and the firm was making a clear attempt to by-pass the union. But perhaps in a less underhand way. Nevertheless, and this is the point I want to stress, subsequent labour relations in Singers seem to have been conditioned more by the first version, which seems to have circulated widely and been believed, than by the second. For some purposes, the fiction captured in oral evidence may be more important than 'the truth'.[42]

Rumours do not survive unless they make sense to people. Seen in such a light, as Portelli puts it, 'there are no "false" oral sources. Once we have checked their factual credibility with all the established criteria of historical philological criticism that apply to every document, the diversity of oral history lies in the fact that "untrue" statements are still psychologically "true", and that these previous "errors" sometimes reveal more than factually accurate accounts . . . The credibility of oral sources is a *different* credibility . . . The importance of oral testimony may often lie not in its adherence to facts but rather in its divergence from them, where imagination, symbolism, desire break in.'[43] History, in short, is not just about events, or structures, or patterns of behaviour, but also about how these are experienced and remembered in the imagination. And one part of history, what people imagined happened, and also what they believe *might* have happened—their imagination of an alternative past, and so an alternative present—may be as crucial as what did happen.

The building of a collective memory can result in a historical force of immense power in its own right; as the epic struggles of the miners, or the repeated persecution of the Jews, or the obstinacy of the Boers, or the three centuries of religious battle in Northern Ireland, so eloquently and tragically testify.

The constructing and telling of both collective and individual memory of the past is an active social process, which demands both skill and art, learning from others, and imaginative power. In it stories are used above all to characterize communities and individuals, and to convey their attitudes. As John Berger has remarked, the function of the stories of past and present which are told in a small community, 'this gossip which in fact is close, oral, daily history', is to define itself and its members. 'Every village's portrait of itself is constructed . . . out of words, spoken and remembered: out of opinions, stories, eyewitness reports, legends, comments and hearsay. And it is a continuous portrait; work on it never stops.'[44] Individual autobiography is less rich in resources. It draws, in a finite span, on what one person has experienced and learnt; and the core of it must be direct experience. But stories are also commonly used in the telling of individual lives, in order to convey values; and it is the symbolic truth they convey, and not the facts of the incident described, which matters most. The encapsulation of earlier attitudes in a story is a protection, which makes them less likely to represent a recent reformulation, and therefore especially good evidence of past values. And this remains so when—as quite commonly in collective oral tradition, and also sometimes in individual life stories—the narrative draws not only on the reconstruction of direct experience, but on older legends and stories. One of my own first interviews was with a Shetlander, born in 1886, Willie Robertson. I asked how much contact the people had with the lairds (the landowners)—a question bearing on their degree of class-consciousness. He told me, as a true story, naming a particular laird, a burial folk-tale which is quite widespread in Scotland:

That was Gifford of Busta. He was one of the county property owners— the laird. And before he died, he'd left instructions that there were to be nobody to attend his funeral except his own kind, the lairds. Well all these people had to come a long distance to funerals and there was no conveyance except they came on horseback. And I have been at a funeral in my time where they give you refreshments: gave you whisky, a glass of whisky, or you could take a glass of wine. Now these lairds that came to Gifford's funeral got refreshments: liquid refreshments; maybe some other. Then they had to carry the remains, the funeral, four or five miles to the cemetery. Well they were always stopping and having more refreshments. And one dropped out; two dropped out; till latterly there were only two; and they lay alongside the coffin. So they were out for the count. And an old crofter come by, and he saw Mr Gifford's remains in the coffin lying there, and these two men. He went across to his house and got a big rope; he took the coffin up on end and put the rope round him; and he took him to the grave and buried him himself. And his kind weren't to be allowed at the funeral. And he buried the laird.

Willie Robertson may have been mistaken in believing his story to be literally true; but this cannot diminish its symbolic force as an answer. Funerals in the island communities of small farmers (crofters) and fishermen were normally occasions for the demonstration of the fundamental equality of all before God, and in the long walk to the cemetery every man would take his turn in bearing the coffin. In some it was even the convention for the better-off to be deliberately paired with the poorest. But as he tells the story he draws not only on a folk tradition, but on his own political and religious ideas. Willie Robertson was an Elder of the Kirk, with a strong belief in temperance. He was also a shoemaker-Socialist: a member of the SDF converted by outdoor speakers who came up north with the East Anglian herring boats. So his story is also a parable of the Good Samaritan; infused with a flavour of Marxist class-consciousness.

Although such a complex instance is relatively unusual in an ordinary life story, it does suggest the need for understanding the different forms and conventions which shape not only stories, but *any* communications between people. Just as in a book the needs of argument, of shape and length, press for the inclusion of some details and the omission of others, so in the telling of an ordinary story: the symbolic meaning and factual details must hang upon a form. 'No utterance whatsoever falls outside a *literary genre*', Vansina urges: study 'form and structure first, because they influence the expression of the content'. Such forms in oral sources have been principally analysed by anthropologists, and by folklorists interested in oral literature, rather than historians. In oral 'literature' distinctions are made between characteristic major genres, such as the group *legend*, the individual *anecdote*, the family *saga*, and the folk-*tale*. Thus there is an international type-list of several hundred folk-tales, which enables archivists all over the world to recognize a tale, and see how the version they have collected varies from the basic type, and what influences have contributed to these changes. Vansina can not only sift out the familiar stereotypes, 'fillers' and 'formulas', from the parts of a narrative which convey significant messages, but also confidently assert, for instance, from the systematic analysis of narratives from a whole region, that 'all migrations in the upper Nile are caused by a quarrel between brothers over an item of little value'.[45] Most European oral historians must work without such a cumulation of experience to help them. The individual anecdote and the family history can be subjected to the same formal analysis—but this has, in practice, been much less often done.

The way in which the story is learnt also needs to be more closely studied. In France, for example, village children are taken by their parents or grandparents to the cemetery to teach them the history of the family. A wedding photograph hung on the wall or a reunion of war veterans or workmates are all mechanisms for the reconstruction of memory. But these mechanisms vary significantly between different social groups and localities. Among the French Protestant minority in the Alpine foothills of Drôme, memory of the past is not of a timeless rhythm of life and work, as for their Catholic neighbours, but of a long, tragic history, a history of struggle and persecution, clandestinity, exodus, and resistance. Here children are shown the secret assembly-places in the woods, the beams from which martyrs were hanged. 'A Protestant—had no right to be born, or to marry, or even to die.'[46] And so deep was the mark of seventeenth- and eighteenth-century persecution, that the more recent past has come to be remembered in just the same mould: the 1851 insurrection not for its marches and clashes, but in its repression; and similarly, the Second World War.

The study of the differing processes of transmission has been carried furthest among the anthropologists and historians of Africa, due to their special dependence on oral sources. A clear distinction must be made between personal oral histories—eyewitness accounts—which are relatively easy to evaluate, and oral traditions—which are handed down by word of mouth to later generations. This latter process can be quite different in two adjacent societies. In northern Ghana Jack Goody has found a sharp contrast between one centralized tribal society, in which a relatively fixed, brief myth is handed down by official utterers; and another society, decentralized, in which performance of the collective myth (the Bagre) is *intended* to be local and creative, so that it continually changes, and different versions from different groups have astonishingly little in common. Other Africanists have tried to disentangle the process by which immediate memory is transformed into formal tradition. This can sometimes be quite rapid: the lives of African prophets, for example, can be transformed into myths within a space of two or three years. But Joseph Miller, on the basis of field-work in Angola, suggests that in some societies there may be a point in time, as events pass beyond the reach of first- and second-hand memory, when recollections undergo a marked change. Accounts of the Angola War of 1861 (which is also known from Portuguese documents) are sometimes relatively accurate, with details of guns and so on, without much moral comment, rather like written documentation; but are sometimes presented as a stylized, mythical event in the traditional war-narrative manner—the style of the official tradition-bearers, the professional oral historians of Angolan society, whose task was to collect oral information and present it in public performances. Possibly memory of the Angolan War is undergoing transition as it passes beyond informal memory. Once none of an audience can remember details of an event, or have their own perceptions and opinions about it, what is needed is a simplified, stylized account which concentrates on the meaning of the story. The time limit thus marks a great sorting-out process, in which some stories are discarded, and others are synthesized, restructured, and stereotyped. The official tradition-bearers are highly concerned with standards of professional accuracy, but they are not the standards of Western historians. Thus beyond that time limit, historians of Angola can best use oral traditions as evidence of values, rather than of facts; and in doing so take account of the completely different African perceptions of time, and of the nature of change, which they embody.[47]

It is hardly surprising, since dates are rarely a strong point even of immediate memory, to find that 'weakness in chronology is one of the greatest limitations of all oral traditions'. Not all is always lost: Paul Irwin has been able to show, for instance, that the sub-Saharan Liptako in Upper Volta correctly remember their Emirs and also the wars of succession between them, at least back to the 1820s. Both here, and in other comparisons from the Pacific between written records and oral tradition, the inaccuracies are far from one-sided. But equally interesting are the distortions: the incorporation of European motifs into traditional history, like the 'wild blackfellow image' of a free past which is now recounted by the North Australian Ngalakan Aborigines, the upholding of false claims to distinctiveness in customs from neighbouring people, the dropping of undesirable rulers from king lists, and the manipulation of genealogies in order to claim land or property which is 'a very common use of genealogies all over the world'.[48]

Above all, consciously and unconsciously, memories which are discreditible, or positively dangerous, are most likely to be quietly buried. 'Forget that story; if we tell it our lineage will be destroyed', exclaimed a Tanzanian in the Nango royal capital at Vugha: his family had a history of conflict with the rulers. Few Germans wish to explore their own family's contribution to the liquidation of the Jews. Even the survivors of such massacres often want to forget, to put the memory behind them, as much as to tell what they had suffered; as Quinto Osano, Fiat metalworker, survivor of the Mauthausen concentration camp, put it, 'Yes, we always want it to be told, but inside us we are trying to forget; right inside, right in the deepest parts of the mind, of the heart. It's instinctive: to try to forget, even when we are getting others to recall it. It's a contradiction, but that's how it is.'

Perhaps this is why the oral traditions of the Australian Ngalaka omit all mention of their decimation by European massacres. Similarly in Turin, stronghold of the Italian working-class movement, the humiliating phase of Fascist domination is typically skipped in factory workers' spontaneous life stories: a self-censoring silence which Luisa Passerini sees as a deep 'scar, a violent annihilation of many years in human lives which bears witness to a profound wound in daily experience'. Lucien Aschieri has shown how, in a less dramatic way, just the same pressures to forget have shaped the local memory of Allauch, a small commune just outside Marseilles, once independent but recently engulfed within the city. Despite their undoubted local impact, the French Revolution, the long local struggle between radicals and the Church, the strikes of 1936, and the two World Wars are given scarcely any attention: because for a threatened community, memory must above all serve to emphasize a sense of common identity, so that episodes of division and conflict slip into oblivion.[49]

Family traditions can be looked at in much the same way. Carolyn Steedman never learnt that her parents were unmarried and she was illegitimate until after her father had died. Jan Vansina, who came from a Belgian village rich in oral tradition, and was first struck by its value when he found the villagers rejected the official version of history taught at school, later found out, after sixteen years of consistent checking, that his own family history was only half reliable. The basic economic story of how his grandfather, in a situation of developing industrialization, went in for growing cauliflowers, is quite correct. But there are more peripheral parts which have either been forgotten as less creditable, or, like the family's distant origins in Milan, created from mis-memories of a visit to north Italy. 'Half these stories are not true. They are an image setting. They are necessary for the pride of someone.'[50]

The discovery of distortion or suppression in a life story is not, it must again be emphasized, purely negative. Even a lie is a form of communication; and it may provide an important clue to the family's psychology and social attitudes. But in order to read these clues, we must develop a sensitivity to the social pressures which bear on them. A spontaneous, unchallenged life story, Vasina argues, will tend to present a consistent self-image or a logical self-development, and events will be 'retained and reordered, reshaped or correctly remembered according to the part they play in the creation of this mental self-portrait. Parts of such a portrait are too intimate or too contradictory ever to be revealed. Others are private but, depending on mood, can be told to the very near or dear. Others are for public consumption.' Thus a typical autobiography may be quite open about family tensions in childhood, but will very rarely reveal difficulties in the writer's own marriage; and less sharply, the same contrast is found when interviewing. Sexual experience is particularly likely to be censored or not told at all. A good interviewer is scarcely likely to be content, however, with the merely public story, and is only likely to find real difficulty in getting beyond this in exceptional cases. One study of deprived families, for example, found that it took several interviews for their informants to move from presenting the answer they thought was socially desirable to one which represented their own views. 'When Elsie Barker was asked how many brothers and sisters she had, her answer in the second interview was that she was the third child out of six.' It was only much later that she explained that the three younger children were in fact the daughters of an elder sister, Brenda, who had committed suicide. Because they were brought up with her, she had always thought of them and continued to think of them as sisters rather than nieces. Elsie had at first 'omitted all mention of Brenda'.[51] The full story was not merely too complicated for a simple answer, but introduced a painful, shameful family memory. Yet it was hidden, rather than irrecoverable. The combination of facts given at different times, through this process of recovery, gives us much more significant information than the bare facts themselves.

The possibility of recovery, of gradually unpeeling the layers of memory and consciousness, is a crucial distinction between direct personal memory and an oral tradition several generations old. There have been a good many anthropologists who have argued that oral traditions are so malleable under social pressures, so continually shaped and reshaped by changing social structures and consciousness, that their value is not only purely symbolic, but that only for the present. Jack Goody, for example, would interpret traditions through a theory of 'dynamic homeostasis', whereby any alteration in social organization or practice is immediately reflected in a reshaped tradition. Vansina vigorously rebuts such extreme functionalism: while it is true that 'all messages have some intent which has to do with the present, otherwise they would not be told in the present and the tradition would die out', the notion that traditions retain no messages at all from the past is an absurd exaggeration. Social changes lead as often to additions, leaving older variations and archaisms intact, as to suppression; and suppressed items usually leave traces. If nothing from the past were left, 'where would social imagination find the stuff to invent from? How does one explain cultural continuities?' What is such functionalism, but the analogy of a machine, mechanically carried to the logical absurdity of a false analogy? And he has still less sympathy for 'the fallacy of structuralism', based on other analogies from unproven universal human laws of thought, and in practice producing an infinite number of interpretations for the same myth: 'despite their pseudological presentation such analyses are in fact creative discourses, valid only to the mind that creates them, aiming at conviction not at proof'. As he cogently sums it up:

Yes, oral traditions are documents *of the present*, because they are told in the present. Yet they also embody a message *from the past* at the same time. One cannot deny either the present or the past in them. To attribute their whole content to the evanescent present as some sociologists do, is to mutilate tradition; it is reductionistic. To ignore the impact of the present as some historians have done, is equally reductionistic. Traditions must always be understood as reflecting both past and present in a single breath.[52]

Such a vigorous defence is scarcely needed for direct personal memory, although the argument would apply here too; the balance of influences is clearly different. Quite often the myth-makers turn out to be not the direct participants, but the reporters, even the historians. The 'classic' southern Spanish anarchist rising in the village of Casa Viejas, seen by Eric Hobsbawm and other historians as a revolutionary response to hunger, 'utopian, mille-narian, apocalyptic', has been shown by Jerome Mintz from direct testimony of the villagers themselves—with whom he lived for three years—to have been a conscientious, but ill-conceived insurrection following the call from Barcelona militants during the 1933 general strikes in the cities. The village had not been a well-organized anarchist stronghold; the rising was brutally sup-pressed before its people had time even to divide up the land, let alone inaugurate a Utopian society; and the man who held out longest was not its charismatic leader, but a heroic and unpolitical charcoal-burner. The myth of Casa Viejas survived because it suited the beliefs of both the Fascist authorities and the Left, providing scapegoats and heroes. And through the Franco decades the survivors had to keep quiet too: 'It's right and natural that not knowing someone well, one would lie. One has to protect oneself.'[53] But they still knew.

For direct memory, the past is much closer than in tradition. For each of us, our way of life, our personality, our conscious-ness, our knowledge are directly built out of our past life experi-ence. Our lives are cumulations of our own pasts, continuous and indivisible. And it would be purely fanciful to suggest that the typical life story could be largely invented. Convincing invention requires a quite exceptional imaginative talent. The historian should confront such direct witness neither with blind faith, nor with arrogant scepticism, but with an understanding of the subtle processes through which all of us perceive, and remember, the world around us and our own part in it. It is only in such a sensitive spirit that we can hope to learn the most from what is told to us.

The historical value of the remembered past rests on three strengths. First, as we have demonstrated, it can and does pro-vide significant and sometimes unique information from the past. Secondly, it can equally convey the individual and collective consciousness which is part and parcel of that very past.

More than that, the living humanity of oral sources gives them a third strength which is unique. For the reflective insights of retrospection are by no means always a disadvantage. It is 'pre-cisely this historical perspective which allows us to assess long-term meaning in history', and we can only object to receiving such retrospective interpretations from others—provided we distinguish them as such—if we want to eject those who lived through history from any part in assessing it. If the study of memory 'teaches us that *all* historical sources are suffused by subjectivity right from the start', the *living* presence of those subjective voices from the past also constrains us in our interpretations, allows us, indeed obliges us, to test them against the opinion of those who will always, in essential ways, know more than ourselves.[54] We simply do not have the liberty to invent which is possible for archaeologists of earlier epochs, or even for historians of the early modern family. We could not have presumed that parents did not suffer deeply from the deaths of their children, just because child death was so ordinary, without asking.

We are dealing, in short, with living sources who, just because they are alive have, unlike inscribed stones or sheaves of paper, the ability to work with us in a two-way process. So far we have concentrated on what we can learn from them. But the telling of their story can also have its impact on them too. We need to turn to that next.

# 5

# Memory and the Self

EVERY historical source derived from human perception is subjective, but only the oral source allows us to challenge that subjectivity: to unpick the layers of memory, dig back into its darknesses, hoping to reach the hidden truth. If that is so, why not seize our chance, unique to us among historians, ease our informants back on to the couch, and, like psychoanalysts, tap their unconscious, draw out the very deepest of their secrets?

It is a beguiling call. Psychoanalysis is the magic of our time. The strange power of psychoanalysts to hear and to heal, to release trapped anger and shame from pasts we had forgotten and, through expression, to put it to rest; to win our love through listening to us and then to give it back to us as a new strength in our own self-confidence—in short, through penetrating to the deepest intimacy that we have shared with anyone, to change our most secret, inner selves—by its nature cannot be fully anticipated or logically comprehended. That alone makes it as threatening as it is compelling. Add a mysterious theory of the unconscious built around our personal sexuality, which is both the taboo and the altar of our culture, and it is no wonder that their power makes pyschoanalysts—and still more psychiatrists, with their battery of drugs in the cupboard to tamper with the mind—the witches, and also the oracles, of the twentieth century. And for historians in particular they present the double challenge, professional as well as personal, of alternative professions manipulating the past according to different rules.

Like it or not, however, few oral historians are going to be able to practice psychoanalysis. It requires years of a different training. Equally important, oral history interviews are based on the assumption of other purposes: our informants cannot be asked to lie on their backs, to open their minds in free association, to talk while the interviewer keeps silent, or to report daily with notes on their dreams and fantasies. But oral historians certainly can learn a good deal from psychoanalysis about the potential of their own craft—both for themselves, and for their informants. Indeed thinking about the implications of psychoanalysis has undoubtedly provided one major stimulus for the advances in our understanding of oral memory as evidence over the past ten years.

The most direct way in which this interest has arisen has often been through a personal experience of therapy. We are fortunate to have one account of this by a leading oral historian, in Ronald Fraser's *In Search of a Past* (1984). This is a rare, original, and fascinating book which would have made a marvellous piece of social history alone. Fraser interviewed his own parent's servants. Through their sharp eyes he reconstructs the Home Counties upper-class social world of the 1930s, and the transformation wrought when the hunting ceased and the social fences came down in the Second World War. The servants' words give us, tellingly, the complex mixture of loyalty and hostility which bound both servant to servant and servant to employer; and also, chillingly, the emotional emptiness at the heart of the manor-house family—the loveless couple, and their lonely, snooty son. But Fraser's courage and originality is to bring together and interweave these painful childhood memories with two other dialogues: with his father, once daunting, now pathetic and bewildered, his mind disintegrated to a blur in which patches of memory float loose, on the way to his end in an old people's home; and in discussion about his own memories with his psychoanalyst. The result is a completely new form of auto-biography, confronting the great issues of time and class, yet intensely intimate.

It is also a fugue on the nature of memory. The apparently straightforward life story evidence of the servants is shown, by juxtaposition, to have its own silences and evasions, for example on the sexual relations between them; it is set against the severely eroded memory of the old man, which may be the fate of their minds too; and it provides material for Fraser's unpeeling of his own unconscious memory in his psychoanalysis. It was his nanny, for instance, who told him how he was fed and changed on the clock, and potted from the age of four months: 'Later, I tied you on your pot to the end of the bed until you produced.' And through his therapy, Fraser not only gives vent to his anger against his parents, but comes to understand how the social division between employers and servants in his childhood home was also an emotional split which he carried into adulthood. His tough, practical nanny was as much a mother to him as his elusive, beautiful mother; while it was the resentful gardener who so hated his father's silent arrogance that became the lonely boy's closest daily companion, listened to him, taught him to plant, to value working with his hands—a second father. It was the gardener who, through emotionally attaching him to a working man, unknowingly opened the political path which Fraser took much later, when he rejected the values of his class along with those of his rejecting father.

Thus while at the start, the psychoanalyst seemed to be looking for something quite different from the past than Fraser's own interest as a historian, brushing aside abstract theorizing, the material world, what really happened, to concentrate on *feelings* about the past, on relationships between people, by the end of his 'voyage of inner discovery' through analysis the two dimensions of understanding had become part of a single interpretation.[1] This does not mean, however, that such an interpretation could only have been reached through psychoanalysis. It would have been an equally typical outcome of the group discussion which takes place in family therapy, drawing out underlying feelings through direct confrontation with other family members in a situation where expressing them is safe, indeed expected. The specific techniques of free association and dream analysis are not part of this therapeutic approach. Nevertheless, it is as effective in uncovering the complexity of contradictory emotions, of intertwined love and anger, which are typical of intimate relationships; and still more so, through the insights brought from a family systems theory which looks for the structuring of relationships, in pointing to the equally characteristic intergenerational influences in emotional patterns.

Take the case of a north Italian small businessman's beautiful teenage daughter, who was slowly starving herself to death. What was her protest about? The family could not understand and the school, where she was working hard and doing well, could offer no clues. Desperate, they came for help to a charismatic family therapist in the big city. Their first accounts of each other were typically restrained: the children thought their mother could perhaps be a bit more independent, while she spoke of her husband as a good man whose only problem was that he never laughed, he was always so serious and sad. But it took scarcely an hour in the consulting room to lift the veil from the family secrets which had paralysed them all. The husband came from a well-to-do family, but had married one of his father's maids after getting her pregnant. For him the affair had been a rebellion against his own father, who dominated his 'saintly', depressed mother; for her, a release from family poverty. But instead of escaping, they trapped themselves in the grandparents' problems, imposing them on their own children. He had done the honourable thing, but he had never forgiven his wife for seducing him and spoiling his life. He preferred to spend his free time with his own parents, sharing a common scorn of 'la serva'. The wife found him always severe, hard with her, unable to listen to her problems, and his scarcely concealed anger had driven her into recurrent depression; the husband found her intolerably over-emotional, was sick of her family's problems; the children complained of her crying and shouting. Emotionally, rather than committed to each other as a couple, both husband and wife remained primarily attached to their families of origin. Her family were not only socially inferior to his, but had remained much poorer, and her bitterest complaint against him was his refusal to give money to help her sisters; while he saw her family as a recurrent drain on his resources, always asking for more. Yet he insisted that each Sunday she must cook a family meal for his own parents who shared his resentment against her and her people. In this deadlock of emotional and class antagonism, the Sunday cooking was a form of hatred. Even though none of them had understood before, it was clear enough now that the girl's refusal of food was a cry against the hidden but intolerable conflict between her parents. Her action was its mirror opposite: rather than food as hatred, starvation as love.

Family therapy is another special situation in which inner truths quite often emerge more quickly than in psychoanalysis. It has the advantage of interpreting individual needs not in isolation, but in a social context. Through its perspective we can explore just why in one family each generation of sons quarrels with its fathers, while another hands down both the skills and ambition to succeed in just the same family profession; why in one family neither fathers nor sons can commit themselves to a single, sustained love relationship, but must always keep mistresses, while in another it is the strong women who call the tune and the men flit through like marginal episodes, and in yet another depressed mothers are followed by depressed daughters. This exploration of the diversity of ordinary experience is far more rewarding than the crude applications of individual psychoanalytic theory to whole cultures which has unfortunately typified 'psychohistory'. It is also much closer to the extraordinary variety in individual lives which oral historians typically discover and need to explain. One of the principal lessons to be drawn from both kinds of therapy is the need for an enhanced historical sensitivity to the power of emotion, of unconscious desire, rejection, and imitation, as an integral part of the structure of ordinary social life and of its transmission from generation to generation.

Similarly, it is not the specific techniques of psychoanalysis in the interpretation of dreams which matter most, but the attention which it has drawn to the pervasiveness of symbolism in our conscious world. We could well ask for dreams from our own informants, their nightmare fears, or their fantasies while day-dreaming on the assembly line; and to learn the most from such expressions of their inner wishes and anxieties we should obviously need to spot the typical tricks of 'dream-work', its condensation of messages, reversals, substitutions, metaphors, word-play, and visual images, through which dreams convey their symbolic messages. These tricks are one reason for the frightening power of fantasy and nightmare. But it is equally rewarding to know that these devices are also normal clues to the symbolic meaning of consciously conveyed messages: of social customs like rough music, or of jokes, or of traditional myths and personal stories.

More directly, the reinterpretation of Freudian psychoanalysis by Jacques Lacan has brought special attention to the fundamental role of language as part of the symbolism. He believes that the unconscious is structured like a language, and he sees the acquisition of sexual and personal identity as a simultaneous and always precarious process, whose foundations are laid as the human infant enters language, through being spoken to, listening, and learning to talk. Masculinity and femininity are therefore imposed on the infant's inner psyche, long before sex differences have any immediate meaning, through the unconscious cultural symbolism of gender embedded in language. Lacan's reformulation of Freud's essentially male perspective on the development of human personality is less radical than those of Klein and Chodorow earlier; and, partly because he has put it forward in such wilfully incomprehensible 'symbolic' language, as a theory it stands up much less well to logical criticism.

Nevertheless, it has undoubtedly helped feminists to show the inadequacies of straightforward deductions from the differences between male and female social achievement, and the hollowness of policies for equal opportunities which ignore the weight of culture. Right from those earliest moments of developing social consciousness, the little girl learns that she is a female entering a culture which privileges masculinity and therefore privileges men, just as in language itself the masculine form always takes priority as the norm and the feminine only enters as the exception. To take a positive place in the world of culture she must fight from the start; but it is an unequal fight. In cultures with pictorial scripts the same lessons will be internalized for a second time, as she learns to read her language: a Chinese girl will discover that the character for a man is made from the symbols of field and strength while that for a woman from loom or womb.

The internalization of such attitudes is equally clearly revealed, as oral historians have found, in the different ways in which older men and women use language. Isabelle Bertaux-Wiame observes how among migrants to Paris from the French countryside, 'the men consider the life they have lived as *their own*', as a series of self-conscious acts, with well-defined goals; and in telling their story they use the active 'I', assuming themselves as the subject of their actions through their very forms of speech. Women, by contrast, talk of their lives typically in terms of relationships, including parts of other life stories in their own; and very often they speak as 'we', or 'one' (*on* in French), symbolizing the relationship which underlies that part of their life: 'we' as 'my parents and us', or as 'my husband and me', or as 'me and my children'.[2] Read in this light, life stories reveal unsuspected and important new messages.

Lastly, we can understand more from what is not said. Again it is not the specific theories of psychoanalysis which prove most useful, so much as a new sensibility, an ability to notice what might have been missed. Freud's own original belief in total memory looks now more like a nineteenth-century fantasy wish to recapture the past, and has certainly no scientific basis, even though it has been so influential that most people apparently 'believe all memories potentially retrievable'.[3] Freud was almost certainly wrong in explaining the absence of memories of infancy through repression: it is much more probable that infant experience is forgotten because the long-term memory is not yet organized than that it is suppressed because shameful. Nor will it help us much to consider whether or not the typical 'resistance' of the analysand—secretive, hidden, obdurate—might be understood through the analogy of childhood refusals to be fed or weaned or to defecate in the right place. The important lesson is to learn to watch for what is not being said, and to consider the meanings of silences. And the simplest meanings are quite likely to be the most convincing.

In short, what we may hope to gain through the influence of psychoanalysis is an acuter ear for the subtleties of memory and communication, rather than the key to a hidden chamber. What is typically repressed is also typically present—such as sex. What the unconscious holds may differ in proportion, and in power, but not in kind: it is simply human experience, accidentally or actively forgotten for all the reasons which we have seen. Concentration camp survivors dream about food and torture. The real world moulds even the delusions of the wholly mad. Victorian schizophrenics wove their fantasies around religion, while contemporary schizophrenics fantasize about sex; but both take off from the everyday concerns of their time. Fantasy and the unconscious are in the end no more than the reordering of lives. Sometimes they may present the world upside down; and they certainly have the power to change how people act in reality. The unconscious is there as a force behind every life story. But the mould of civilization and its discontents is clear enough, from whichever side of consciousness we perceive it.

There is however another dimension of psychoanalysis which demands equal attention from oral historians. This is its primary therapeutic process through the releasing of memory. Many oral historians have come to realize this by chance, through their own practice. They will learn—often through a third person—how being interviewed gave an old person a new sense of importance and purpose, something to look forward to, even the strength to fight off an illness and win a new lease of life. They may also have found that it is not always so simple.

Most people hold some memories which, when recalled, release powerful feelings. Talking about a lost mother or father may evoke tears, or anger. Usually an unembarrassed, sympathetic response is all that is needed in this situation: expressing the feelings will in itself have been positive. But some memories tap deep, unresolved pain which really demands more sustained reflection with the help of a professional therapist; and clearly in such cases the best that the oral historian can do is to suggest how this could be found. They arise, typically, from family experiences which are violent, shameful, or especially entangled and perplexing; or from the traumas of war and persecution.

Donald and Lorna Miller have recorded the memories of survivors of 'the first genocide of the 20th century', the massacre of over a million Armenians—half their entire population in the Ottoman Empire—in the years from 1915 to 1922. Some were burnt alive; others used as pack animals, maimed, tortured, or starved to death. The bellies of pregnant women were torn open, and other mothers forced to leave their straggling infants to die on rocks in streams, or sell them to passing Arabs; not a few mothers and families committed suicide together. What has this unimaginable horror left in the memory of survivors? Some will never speak of it. In some, anger has now given way to political explanation, or resignation that nobody cares; or even to forgiveness. But in others hatred of the Turks burns fiercely on: they dream at night about being chased—'I wake up sweating'—'They stab you in the back.' And yet others still hope for revenge: in 1973 one 78-year-old survivor, who had lost almost all his family in the massacre, shot two Turkish consular officials dead in a café in Santa Barbara, California.[4] His rage was a memory which spanned over fifty years in another culture and another continent.

The memory of the still more systematic degradation, humiliation, and extermination of the Nazi concentration camp victims is equally haunting. Two hundred Italian survivors have testified how many of them kept it to themselves, because they felt its full horror would be incredible to others, inexpressible in words, and too painful for those close to them to hear: how they had been separated from all they knew, robbed of all their possessions, stripped naked, shaven of all their hair, given numbers instead of names, made to eat with their mouth and hands 'come una bestia'; lived every day within the sight and smell of death, smelling the burning bodies, seeing the human ashes used for road cinders, seeing piles of corpses; how they learnt, in order to survive, to eat grass to keep down hunger, to steal from anybody, to trust nobody but an intimate, to sleep undisturbed next to a corpse of a fellow inmate after stripping it of clothes to keep warm themselves, above all to think of death as ordinary, even when guards hacked another prisoner to death in front of them by beating their heads open . . . No wonder even today, the price of the telling may be weeks of renewed nightmare terror. And such memories may be almost as intolerable indirectly. Claudine Vegh found similar fears, nightmares, nervousness, anger, and paralysis when interviewing French Jews whose parents were killed in the Nazi period. 'Many of those who were left orphans never talk about their past, it is taboo . . . They do not want to, above all, they cannot, talk about it.' Many who did talk were very reluctant, spoke in hoarse whispers, or burst into tears. They had not been able to mourn at the time of their separation from their parents because there had been no time, no ceremony, and they had not been sure until long after that the parent had died. No reconciliation had been possible. They had carried into old age an open wound, a confusion of loss, shame, anger, and guilt, still as real today as ever: 'mute agonies' which had 'haunted them all their lives, . . . a hurt so painful, so omnipresent, so all-encompassing, that it seems impossible to talk about it even a lifetime later'.[5]

Such memories are as threatening as they are important, and demand very special skills in the listener. They are thankfully exceptional. For most people the pain of the past is much more manageable, lying alongside good memories of fun, affection, and achievement, and recollecting both can be positive. Remembering our own lives is crucial to our sense of self; working on that memory can strengthen, or recapture, self-confidence. The therapeutic dimension of life story work has been a repeated discovery. Thus Arthur Ponsonby, the literary critic and anthologist of *English Diaries* (1923), noted how many of his authors used their diary pages for the purpose of 'self-analysis, self-dissection, introspection, . . . for clearing their minds, for threshing out human problems, for taking stock of the situation . . . They may even derive from it the same sort of relief as others find in prayer.' Willa Baum has found that oral history interviewing 'almost always does have beneficial effects for the narrators'. Sociologists have also noted the confessional dimension to life story interviewing, and partly because much of their work has been with deviants who are often personally isolated, have especially encountered unexpectedly warm responses to a 'sympathetic ear'. Annabel Faraday and Ken Plummer vividly illustrate this from one series of letters they received:

If my reactions have been impulsive, it is because you have unexpectedly breached the wall of my isolation and I cannot help thinking of you as a friend in a special category. Hoping you can think of me in a similar way.

And from a later letter:

I found great relief in talking to you today. Thank you for being such a sympathetic audience and for making me feel so relaxed.

And again, several months later:

I feel I am overburdening you and using you as an outlet for my personal troubles, but it has been a case of opening a valve . . .

As researchers, they did indeed find their shift from sympathetic observer, 'through sounding board to confessor and emotional prop', was a burden which could have consumed indefinite energies, given the severe problems of many of the sexual deviants they were recording. But the positive changes which they saw in some informants were equally striking: the transvestite, for example, who suggested he was now 'strong enough to "come out" publicly—a move which he felt would inevitably result in the final breakdown of his shaky marriage and which he suggested could be done through the publication of his life history'.[6] The changes which oral historians may notice in their subjects are certainly unlikely to be as colourful, but they may be as important. Their growing realization that not only were people good for history, but history might be good for people in themselves, has been one major source for the reminiscence therapy movement which has spread so strikingly in recent years.

Its other growing point has been a noticeable change in attitudes to ageing in the caring professions. Twenty years ago gerontologists actively disapproved of reminiscence. They saw 'living in the past' as pathological, a withdrawal from present reality, denial of the passage of time and ageing, even evidence of organic brain damage or psychological 'regression to the dependency of the child'. The idea that reflection on the personal past, and through it acceptance of change, might be essential to the maintenance of self-identity through the typical transformations of the life cycle, is a logical inference of basic psychoanalytic thinking, and was already being argued by Erikson; but psychoanalysis of older people was in practice uncommon. The key influence came from an American research psychiatrist, Robert Butler, who began tape-recorded interviews to investigate the mental health of the old in 1955, and through interviewing stumbled on the 'quite apparent . . . therapeutic benefit in reminiscence'. Beginning with a now classic paper on 'The Life Review' (1963), he has argued for the need to view reminiscence as normal and healthy, part of a universal process for re-evaluating past conflicts to re-establish self-identity, and a means of helping the elderly to help themselves. Through either an individual interview or a group discussion 'they can reflect upon their lives with the intent of resolving, reorganizing, and reintegrating what is troubling or preoccupying them'. The old as much as the young need the chance to express their feelings, talk through their troubles, work out their regrets; to reconsider, for example, at a time in life when they want to hand on their moral experience to a younger generation, the painful lessons of parenting, and 'to express the guilt, grief, uncertainty, fear, and uneasiness which are tied to their concern that they have not been effective parents'.[7]

Butler's influence was spread both through a guide to *Living History* (1970) published by the American Psychiatric Association and also a series of papers by researchers and therapists attempting to test his ideas. Partly because of disagreement over what exactly was meant by normal reminiscence, and also through insufficient consideration of the distorting effects of loss of self-identity through institutionalization, nothing sufficiently conclusive emerged from this research.[8] The eventual practical breakthrough in Britain has come rather from the social workers and hospital staff who have provided the front line in caring for the rapidly growing numbers of older people in need today.

There were already some precedents in social work practice, such as the use of 'life books' of documents and photographs first developed for children in care, to help them keep or recover a clear sense of self after bewildering transfers between institutions or while settling in with foster-parents. Life books have been more recently used with the adult mentally handicapped. There were also instances of the casual use of reminiscence in social work with the elderly. The crucial point, however, was a growing realization of 'the huge arrogance', as Malcolm Johnson has termed it, of professionals—of a different class and generation, and a different life experience— presuming that they could define the needs of their clients without first understanding their own diagnosis of their condition. That meant looking at the problems of old age in the perspective of the ageing person's own life experience: listening to them, 'to identify the path of their life history and the way it has sculpted their present problems and concerns'. In this way the individual's own *priorities* for the later phases of life would emerge, the outcome of a lifetime of 'losses and triumphs and fears and satisfactions and unfulfilled aspirations'. Many social workers stumbled on this relevance of the past when it proved the clue to a particularly puzzling case: the old man, for instance, unable to look after himself, yet stubbornly reluctant to go into a home. His resistance became immediately understandable once it was learnt that he had been a child in an orphanage. Listening, in short, was professionally useful.[9]

In the mean time a government architect for old people's accommodation, Mick Kemp, disturbed by the low quality of life he had seen in the many homes he had to visit, had been enabled to develop experiments in the use of images to stimulate withdrawn old people to talk and respond. The first images were artistic, but then he found that old pictures of scenes and events—like domestic interiors, the abdication of Edward VIII, the General Strike, or the Jarrow March—were still more effective. So was old music, linked to the pictures. From 1980 the *Recall* project was taken over by Help the Aged Education Department, where Joanna Bornat was able to bring to it her own experience as an oral historian. Within a year a series of six tape and slide sequences combining music, singing, and spoken memories of the past was issued for practical use with groups of old people.

Cheap and simple to use, needing only a tape recorder, a slide projector, and a white wall to get started, the *Recall* packs have proved immensely successful, an immediate and practical demonstration of the effectiveness of Butler's ideas. In a whole variety of situations, ranging from old people's day centres and clubs for the active elderly living at home to the chronic sick in danger of institutionalization, the severely depressed, dementing, or even psychotic hospital patients, reminiscence has been reported as having remarkable effects. In a normal group of rather bored, withdrawn old people, there will be a sudden change of atmosphere: as the tape-slide show goes on, they start to talk, and to sing the songs, and then they go on talking afterwards—and the others listen to them. Still more remarkably, old people who have been mute for months suddenly speak to others; someone normally totally inert can be seen tapping a foot to music; an old woman who has seriously neglected herself starts to take an interest in her appearance; an old man, silently withdrawn into acute depression, exchanges a memory with a former workmate—and he smiles. Equally important is the impact of this changing atmosphere on others. Staff can see that something can be done; relatives begin to visit more often, and stay longer. Put simply, *Recall* sparks a common talking-point; and once communication is restarted, people rediscover each other as human beings.

The first series of *Recall* packs—on 'Childhood' in the 1900s, 'Youth' in the Twenties, 'The Thirties', the two wars, and so on—sold 1,500 in three years. Several districts have now produced their own local packs. There are regular training courses in the use of reminiscence for nurses, social workers, occupational therapists, and other professional and voluntary workers with the old. Reminiscence therapy has become a reinvigorating wave, a catalyst for change in the care of the older people which can alter attitudes in a host of small but cumulatively significant ways. John Adams put up old posters in a continuing care ward in a south London hospital and found these too got both residents and visitors talking: 'patients can be overheard explaining to their visitors, with the aid of an Imperial War Museum poster, how they distinguished German from British airships in World War One.' Wards should be given names which mean something to old people, after a Music Hall star, rather than after a saint or a local landowner, and they should be decorated with the same intention, so that 'reminiscence ceases to be a special event and becomes simply part of the general fabric of ward life'. Andrew Norris and Mohammed Abu El Eileh took out a group of frail elderly psychogeriatric patients from a Dartford hospital, following a suggestion in a group therapy session, on a tour of places of interest which they had mentioned, visiting former workplaces, homes, schools, and pubs: 'the response of our reminiscers was staggering.' Group discussion also resulted in the discovery of three former musicians among the patients, for whom they were able to buy instruments to entertain the ward. More active groups have gone on to produce local booklets and exhibitions. Sometimes work of this kind, such as the life story of a White woman married to a London Chinatown opium dealer in the 1920s and later a dealer herself, recorded with Annie Lai by Bob and Pippa Little in a Hackney old people's home, has proved not only to restore the confidence of an old woman ashamed of her past, but also to be uniquely valuable historical evidence.[10]

It is still far from clear how useful, in a strictly medical sense, reminiscence may be. Certainly it cannot cure conditions like dementia: 'it may just make living in hospital a little more bearable and meaningful.' Equally clearly its application and its effects must differ, depending on whether the context is social work with old people living in their own or residential homes, or in a range of hospital situations; and between individuals. Peter Coleman emphasizes that reminiscence does not suit everyone equally. His own research is exceptional, for he has been able to follow up the eight survivors of a group of fifty-one old people whom he had interviewed in sheltered housing in London ten years earlier. Originally he had found that twenty-one were 'happy reminiscers' who enjoyed talking about their pasts; but there were also sixteen who saw no point in reminiscing, because they were actively coping with life in other ways. The 'happy reminiscers' proved the most resilient group, but the active non-reminiscers only included a few with low morale, such as a lonely ex-prisoner: more typically, they were busily active and reminiscence seemed to them a waste of precious time. The more striking contrast was with two other groups. Eight were 'compulsive reminiscers' whose 'brooding in the past' was 'dominated by regretful memories': they talked a lot, but felt bad about it. Their prognoses proved bad, showing increased psychological disturbance. Such people might actually be harmed by group therapy: what they needed was skilled personal counselling. The prospects were equally bleak for the six who avoided reminiscing because it made them more depressed, since the present seemed to them so much worse than the past. Typically they had suffered a severe loss, such as recent bereavement, and could not manage the difficult adjustment to widowed, single life without their lifetime companion. Again their need was for individual therapy, which they did not receive: those who had not died were still depressed ten years on. There are, in short, no automatic solutions: 'each person needs to be considered in a special way.'[11]

The key is in precisely that. Reminiscence therapy is no more a panacea than psychoanalysis. The basis of its transforming power, as with oral history itself, is quite simple: listening seriously to what old people have to say. It is through this that a miserable, difficult, moaning old White person can become a whole person, even one with some similar experience, to a young West Indian nurse. When care staff do not get to know their patients they too easily become mere bodies to be fed and watered, managed and maintained in existence. Communication can make them people again.

> What do you see nurses, what do you see,
> What are you thinking when you look at me?
> A crabbit old woman, not very wise,
> Uncertain of habits with far-away eyes,
> Who dribbles her food, and makes no reply
> When you say in a loud voice, 'I do wish you'd try' . . .
> Is that what you are thinking, is that what you see?
> Then open your eyes, you're not looking at me . . .
> I'm a small child of ten with a father and mother,
> Brothers and sisters who love one another.
> A young girl at sixteen with wings at her feet,
> Dreaming that soon now a lover she'll meet . . .
> At twenty-five now I have young of my own,
> Who need me to build a secure happy home . . .
> At forty my young now will soon be gone
> But my man stays beside me to see I don't mourn . . .
> Dark days are upon me, my husband is dead:
> I look at the future, I shudder with dread . . .
> I'm an old woman now and nature is cruel,
> Tis her jest to make old age look like a fool . . .
> But inside this old carcase a young girl still dwells,
> And now and again my battered heart swells,
> I remember the jobs, I remember the pain,
> And I'm loving and living life over again . . .
> So open your eyes nurses, open and see,
> Not a crabbit old woman, look closer—see *me*.[12]

# 6

# Projects

ORAL history is peculiarly suited to project work. This is because the essential nature of the method is itself both creative and co-operative. It is true that oral evidence once collected can be used by the traditional independent scholar who works only in the library. But this is to miss one of the key advantages of the method—the ability to locate new evidence exactly where it is wanted, through going out into the field. And field-work to be successful demands human and social skills in working with informants, as well as professional knowledge. This means that oral history projects of any kind start with unusual advantages. They demand a range of skills which will not be monopolized by those who are older, expert, or best at writing, so they allow co-operation on a much more equal basis. They can bring not only intellectual stimulation, but sometimes, through entering into the lives of others, a deep and moving human experience. And they can be carried out anywhere—for any community of people carries within it a many-sided history of work, family life, and social relationships waiting to be drawn out.

Oral history projects can take place in many different contexts, both as individual and as group enterprises: in schools, colleges, and universities; or in adult education or literacy groups, from museums, or from community centres. They can involve all kinds of people: schoolchildren, unemployed young adults, working parents, or the old retired. Although sharing many features, each context provides a distinctive emphasis which carries its own advantages and its own problems. We shall therefore look at them in turn. And because the discussion is meant to be a practical one, we shall recount the particular experience of several projects which have been carried out recently. The ideas which follow are not just ideal suggestions, but have been shown to work.

Let us begin with schools. Since an oral history project is a complex and time-consuming operation, the first question any teacher is likely to ask is why give it a place in the curriculum at all? The educational arguments can be summarized briefly. A concrete objective and a direct product are provided for project work. Discussion and co-operation are promoted. Children are helped to develop their language skills, a sense of evidence, their social awareness, and mechanical aptitudes. For history teachers oral history projects have the special advantage of opening up locally relevant history for exploration. But they have also been successfully used in teaching English, social studies, environmental studies, geography, or integrated studies; and, in varying forms, at any stage of social and intellectual development between the ages of 5 and 18.

Any school project in oral history should assist children towards a much sharper appreciation of the nature of evidence, because they will be directly involved in its collection. This may come as a revolutionary and undesirable idea to authors and publishers who dislike school projects which create their own resources, or indeed to some professional historians. But at a simple level, in collecting anecdotes and memories of how people lived in the past, how they dressed, children's games, the changing landscape—however primitive their interviewing or recording technique may be—children are collecting evidence. At the same time, they become creatively involved in assessing it. They face basic issues: when to trust or to doubt information, or how to organize a set of facts. They experience, at a practical level, history as a process in the re-creation of the past. Like young archaeologists, they are given spades in the place of lectures— taken to the coalface to hew as historical researchers. And because they are collecting from sources which professionals have not used, they have a chance of putting this new evidence together in a piece of history of their own—a great-grandfather's memories of the First World War, or a neighbour's memories of a street fifty years ago—which is unique, and so brings a very special sense of achievement.

During this process, several types of skill may be developed. First, there are inquiry skills. Once pupils have started to interview, the desire to find out more from other sources can be very powerful, leading into searches of the school or local libraries for books, and through this into techniques like using book indexes or the library cataloguing system. They learn through a range of techniques, not just by interviewing.

Next, it can provide important assistance in the development of language skills. This concerns both written and spoken language. Before interviewing, children have to discuss together the best wording of the questions which they might ask. When tapes are played back, they can also criticize the way in which questions are asked. While themselves interviewing, they have to learn to listen to other people, and grasp exactly what they intend to convey. This can demand intense concentration. Without realizing it, they are confronted with the problems of comprehension and interpretation which the English textbook comprehension exercise attempts to simulate. At the same time, through interviewing, or through being themselves interviewed, children can gain confidence in expressing themselves in words. This can be transferred from the spoken to the written word, for example, by getting them to write down what they can hear from a tape; or, conversely, by using a duplicated version of a transcript as a starting-point for discussions. They could perhaps discuss the differences between written and spoken language. At a later stage in the project, they can go on to reading in library work, and to a presentation of the project, including transcript material, in written form.

Technical skills will also be acquired in the handling of machinery when tape recorders are used—although these are definitely not essential in school projects. These skills can be carried further in the presentation of the project: for example, by editing extracts from recordings into a tape sequence, or by combinations of slides and sound, or by printing a booklet combining photographs and transcripts, or by an exhibition which makes use of all these means together.

Finally, fundamental social skills can be learnt. Through interviewing itself, children may develop some of the tact and patience, the ability to communicate, to listen to others and to make them feel at ease, which is needed to secure information. To interview you have to behave as an adult; you cannot giggle.

Children can be helped to learn to move in an adult world. At the same time, they may gain not only a vivid glimpse of how life was in the past, but a deeper understanding of what it is like to be somebody else; and how other people's experience, in the past and today, is different from their own—and why this might be. They can thus be helped both to understand and feel empathy with others, and to face conflicting values and attitudes to life.

So much for the theory. What about the practice? We can best turn to some examples of projects which have worked. The first is from a primary school.

At a county primary school in Cambridge, Sallie Purkis has used oral history with the younger age groups. She began with a project, carried out over half a term on two afternoons a week, with a class of twenty 7-year-olds. It was a diverse group: some of the children came from abroad, and while nine children could not read, others were very bright. The project was to be their first ever experience in learning history. One of its objectives was to make this first encounter exciting and interesting, and to get the children to feel that they could collect historical evidence, and that history was real, and relevant to their own present. It was an advantage that the project was carried out in a school without subject boundaries, so that she could launch easily into art work, English, and outside visits.

She chose as a concrete starting-point a photograph, suggested by a local librarian, of the school itself sixty years earlier, just opened, with its first pupils standing among the builders' rubble. The children were immediately interested by this, commenting on the pupils' clothes. They worked out where the photograph had been taken from, and how old these first children would not be—in other words, as old as their own grandparents. Following this, 'grandma' was chosen as the key symbolic figure of the project (aunts or other relatives could be substituted)—and it turned out that it was a novel experience for *grand*parents to be involved in the school. Tape recorders were not used, but a written questionnaire was sent out. It was composed after discussion with the children, and, in retrospect, was too long, for it produced more material than could be organized satisfactorily—a few questions would have been quite sufficient. Most but not all of the grandmas responded, and one child, who called himself a 'historian' by the end of the project, interviewed three people. Another produced a typescript. There was thus an abundance of good material.

Sallie Purkis made a reading book for the class by selecting extracts on particular topics and writing them out herself. The first topic was 'What Grandma Said About Clothes'—men's clothes, women's clothes, and shoes—one child's grandfather was a shoemaker. The children drew these. They also brought in photographs, often very precious, so that they had to be protected in plastic; these made a big show, and the children proudly identified with them. Then objects began to be brought in—garments, irons, and so on. Some of them were rather overwhelming, like 'the hat my father wore at grandfather's funeral', in a big box marked NOT TO BE OPENED. Some of the children went on to reading (although it was difficult to find suitable books for this age group). Other children made a model clothes shop out of shoeboxes. The class went on a museum visit. All the children wrote essays—on shopping for clothes, on washing day; and on 'Grandma Day'. For the climax of the project without doubt was Grandma Day: the afternoon when, to their own very apparent enjoyment, the grandmas were invited up to school for talk and tea with the children.

Another imaginative primary school project—one of many regularly reported in the news columns of *Oral History*—was at an inner London school in Notting Hill. For this Alistair Ross chose a history of Fox Primary School itself, focusing particularly on the school's wartime evacuation to the countryside. This enabled the project to include a school journey with the two classes of seven- to ten-year-olds to Lacock in Wiltshire, where they interviewed villagers who remembered the invasion of London children from their own school over thirty years earlier. The whole project was notably successful in stimulating the children to develop collaborative discussion with adults, and also to use what they learnt in their own creative writing.

Next, let us look at some projects with older children. In this context both the constraints and the opportunities are heightened. The interdisciplinary and investigative nature of oral history requires some flexibility in the school's timetabling, and also the adoption of the project-based examination schemes now available, for realizing its full potential: but given this scope, a project with older children can reach more ambitious levels. Many schools have been able to publish projects for their own centenaries or on local history as booklets; others have organized local exhibitions. At Thurston Upper School in Suffolk, Liz Cleaver used some form of oral history with all the years: constructing family trees with the younger classes, and then moving on to comparisons of health provision before and after the welfare state by questioning grandparents, and finally to a series of sixth-form projects on 'Life in Suffolk Between the Wars'. The sixth formers set up a committee to run these projects themselves and used them to produce special magazines for local sale. Each year a minority went on to individual A-level projects for their school-leaving examinations.

Nor is it just the brighter children who have gained from project work. In a Manchester comprehensive school, Ruth Frow has used oral history for fourth-year children. This was a remedial class who were taking a C.S.E. local history course partly assessed on project work. They did not respond to demonstrations by experts of how to interview, but were much more interested when a voluble old lady, one of the pioneer settlers of the district, was brought in to talk to them—although even so they remained virtually tongue-tied. Afterwards, however, they all worked hard on the project. One carried out a tape-recorded interview with his grandmother, asking the questions which they had worked out in class discussion. And the school truants in the class insisted on continuing with the project even when suspended.

She also played the tape from this project to a first-year mixed-ability class of eleven-year-olds. This resulted in an unexpectedly striking success story. One pupil, John Macdonald, was so fired with enthusiasm that he started his own 'Research in Old Peoples Home'. They drew up a list of questions together and he successfully wrote down the life story of one eighty-five-year-old resident. He then went on to tape recording. He did well with his own grandmother, but he failed spectacularly with a stone-deaf 102-year-old ex-Chinese missionary, despite a heroic persistence in repeating the same question time and again. And the project came to a sudden end when John Macdonald formed a group and asked to interview Ruth Frow herself, explaining that they intended to interview *all* 'old' people.

A special form of life story work has also been developed for remedial teaching in reading and writing, both in schools and also for adult literacy groups. For many of the illiterate or scarcely literate adults—now estimated at two million in Britain—and children, shame and secrecy at their disability are the first hurdle which needs to be overcome if they are to be helped. Often they face the added disadvantage of being first-generation immigrants, attempting to communicate in a second language. Beginning from oral work enables them to start where they are most confident and build from that point. Typically a group will first discuss some personal experiences of common interest: family, work, migration, or whatever. The tutor will record this discussion and transcribe parts of it. At the next session students will attempt to read part of the transcript of their own words, and then discuss the ideas, words, and grammar they have used, with the tutor suggesting possible changes to make the passage clearer. Gradually they move on from reading what they have spoken to writing independently, and to reading generally. But throughout the clue to success is beginning with their own language, as it is spoken, with its personal phrases and speech rhythms, about familiar people and experiences. It is this which eases each leap forward. Often the end result may be a personal autobiography, and some of these have later been published in booklets like the London collection *Our Lives*, which included the migration story of Mohammed Elbaja, a Moroccan boy at Shoreditch School. Many more are able to see their own words printed in the magazine *Write First Time*, which has developed its own special format, with the lines broken to reflect points of sense and rhythmic phrases within each sentence, as in the example shown in Figure 1. With a circulation of some 6,000 copies per issue, it serves a vital educational movement closely linked with the experience of oral history.

Another specialized form of educational work is reminiscence drama. This is again practised with various age groups, and there are companies like Age Exchange who create theatre from old

This man said I was to slow
For Him ~~so I~~ at Work. So I told
the DRO Lady about Him and
He threatend with the sack he
He sent my mother and dad a
letter so we went see the disabled
Lady and she telephond him
and she was ver angry abuot
it ~~I was~~ we were pushed
out of the door lake criminal
and could not ~~se to~~ say ~~a~~ anything
about it. but ~~...~~ acording to
His set His time and ~~..~~ emotion
man. He made redundant

Figure 1 This short piece by Terry Collins, which appeared in *Write First Time*, was written as an angry response to being made redundant. The first idea, which was two sentences long, was typed up and duplicated. Then everybody in the group discussed the issue and Terry added to it. Above is his final draft. He was then given some advice on spelling and setting before the piece appeared in the form printed below.[1]

## Made redundant

by T. A. P. Collins

This man at work
told the boss that I was too slow for him.
He sent my mother and dad a letter
so we went to see the Disabled Resettlement Officer.
I told her that he threatened me with the sack.
She telephoned him as she was very angry about it.

Me and this other boy, we were pushed around,
out of the door like criminals
and could not say anything about it.
But according to him, the time and motion man,
I was too slow.

He made me redundant.
I had been there eleven years and nine months.

people's memories which they play back primarily for audiences of old people, in clubs, community centres, or homes. But the new Royal Court Young People's Theatre by the Portobello market works especially with London schools, and its director Elyse Dodgson draws on her own remarkable experience in creating and producing theatre with West Indian girls in an inner south London comprehensive school. Three of their plays reached the London stage, culminating in *Motherland*, a deeply moving but beautifully controlled drama of West Indian experience—hopes and dreams, reality and rejection—in coming as migrants to Britain. The starting point was a set of interviews with the girls' mothers and others of their generation in Brixton, collected by a former pupil with the help of a special grant. This personal testimony fired the girls' imagination and sustained them through the long period of working together as a group, for six hours a week over several months, to create the play. The emphasis was on group work rather than individual performance, with all expected to participate in every rehearsal, each playing a variety of roles, and all decisions taken through mutual discussion. The drama was developed through experimental role play: responding to themes taken from the recorded testimonies, the girls developed expressive mimes and wrote the words and music of songs to match the individual words. At a later stage the mothers were drawn in to see rehearsals and make their own suggestions. Ultimately *Motherland* interwove three levels of expression: the remembered real experience of the text from the interviews, spoken by narrator or chorus, the children's own imaginative songs, like 'Searchin' for housing, and the symbolism of group mime. Thus the experience of asking a landlady for accommodation and being turned away was expressed through a ritual of the whole group, walking, knocking, and freezing, which became a central image of the whole play. It is this combination of creative but highly disciplined group expression with the words of individual life experience which makes this children's drama so unusually compelling.

It is, however, to the United States that we must look for the most widespread use of oral history in secondary schools: for there is still nothing in Europe to approach the extraordinary success of the Foxfire project. Eliot Wigginton—seeming just an ordinary teacher still, although he must have some extraordinary skill—went straight from college to Nacoochee School, a small town high school at Rabun Gap in the mountains of Georgia. He soon discovered that the teaching methods he had been taught simply did not work. In the teachers' room the talk was utterly pessimistic: these kids could not do anything, they could never write. Wigginton saw that their essential problem was *boredom*: they were 'just ordinary high school kids' with a hunger to *do* something. Once they got the chance to create their own material they were transformed. They found it in the *Foxfire* school magazine project. Originally started as an optional course in 'Creative Writing', it is now the core of a cluster of courses which allow credits in filming, building and other crafts, and even science. The children collect information through interviews, photography, and measured drawings, and they learn not only to produce a magazine, but technical crafts, taking apart and reconstructing machines and old buildings, and organizing old people's get-togethers.

The magazine has been a phenomenal success, not only locally, but nationally as it is now known right across the United States. Selections published in book form from 1972 as *Foxfire One, Two*, and *Three* had already sold more than four million copies by 1978, with the first volume proving the best-selling book ever published by Doubleday. Partly through its own foundation with its regular training workshops and newsletter *Hands On*, the influence of *Foxfire* has spawned more than two hundred similar high school magazines, such as *Sea Chest*, published by the Cape Hatteras High School at Buxton, North Carolina, and *Loblolly*, from Gary High School in the ranching-lumbering, gas, and oil countryside of Texas.

Wigginton is in constant demand as an adviser all over the country, especially from demoralized inner city high schools. I heard him describe a typical visit to an Indianapolis High School with 5,000 pupils and 350 teachers who had 'just given up'; only the twenty school guards, armed with guns, maintained a bare semblance of order. His message there, as always, was that responsibility must be placed, not on the children, but on the teachers. If children were given the chance to be creative, they would respond. He would abolish textbooks altogether if he could.

Certainly the school's location has helped to create *Foxfire*'s wider appeal. Over the years, the project has built up a group of trusted informants, to whom the school pupils return again and again for new information. Their southern voices speak of regional crafts—log-cabin building, cooking, making quilts, soap, straw mattresses ('Best sweet smellin' bed you ever slept on. We changed th'straw every year at threshin' time')—and of customs and beliefs around childbirth, work, sickness and death, which take one straight back into the world of the early Puritans, in which religion and witchcraft were still inextricably intertwined. Of curing burns and bleeding by faith healing they say, 'Stoppin' blood's just like drawin' out fire . . . You do it with th'same verse and th'same words.' And of planting by the moon:

Take taters. On th'dark of th'moon or th'old of th'moon—that's th'last quarter, they make less vine; and on th'light of th'moon they makes more vine and less tater . . . Th'Lord put th'signs here for us t'go by. It's all in th'Bible: th'signs of th'stars, moon, sun, and all. You've got to follow all these signs if you do right. Don't you know the signs?

Yet to get schoolchildren to bring together material of this quality in a regular quarterly magazine, even after winning the confidence of the people of old mountain communities, has taken great organizing skill and imagination over several years—especially when it remains, as any oral history project must, just one aspect of a school's programmes, exciting some pupils, but by no means all.

Eliot Wigginton starts with a relatively brief preparation for interviewing in the classroom:

We illustrate, for example, the need for a tape recorder by telling a short story in the class, and then the next day without telling them the reason for telling the story at all come into class and say, 'Let's take out a piece of paper and a pencil and retell that story that I told you yesterday.' I use a few details; say it's a hunting story, I can say, 'When I caught that bear I took a stick about six-and-half feet long, and I did this with it, and then I took a couple of rawhide thongs and I did this and I . . .'—that kind of thing. I tell a story that's got some details in it that a kid would be forced to remember. You'd be amazed how off the kids are in their repetitions of that story.

This leads on to a discussion of how to get a good interview. The key is genuine curiosity:

The kid should not be wandering around and looking out of the window and drumming on the floor . . . You may get into an interview situation that you expected to take a half an hour and it may in fact go into four hours if somebody gets cranking. They've got to be curious . . .

You tell them that the cardinal sin of interviewing is to get into a question-and-answer pattern . . . get one sentence or two sentence responses and then ask another question on a totally different subject . . . What you want an informant to do is to get on to a topic and then begin to expand, and inside that expansion all kinds of things happen. You try to get the kids to ask the same question a hundred different ways. You know, 'How did you do so-and-such? Well, did anybody else in your house do it differently?' You keep them beating around inside that topic as much as possible—'Have you ever heard of it being done in another way?' Then if possible you give the kids some information before they go out on a topic that they can carry with them, like other alternative means of doing something . . . (And) you have to remind the kid that he can't cut off the people once they get going. If they start telling stories . . . bear hunting stories, the worst thing you can train a kid to do is cut him off and say, 'No, wait a minute! I don't want bear stories. I want how to tan a hide . . .'

This preparatory stage is kept quite short. As soon as possible pupils are sent out into the field for their first interview. They go for the first time either with a staff member, or another more senior and experienced pupil. Initial interviews are set up in advance, to make sure that the experience is with a sympathetic informant. Once the children become more confident, they go out into the field on their own without any staff members. And they begin with relatively small topics—'how one woman made soap, smaller things of that sort'—and gradually work up to bigger topics.

Eliot Wigginton's most remarkable achievement, however, is in devising an organization for publication which makes the fullest possible use of the material which the pupils collect:

We do many kinds of articles. I'll give you a few illustrations. We got a call from another magazine one time that said, 'We're doing an article on various ways that you can tell what the weather is going to be like this year . . .' We had six or eight in a file, and we said, 'Okay, I'll tell you what. You call us back at 3.30 this afternoon and we'll have something for you.' You take two class periods and just about every kid who is working on the magazine—fifty-five kids—and you say, 'Okay, you guys right there, you take Kelly's Creek. You guys take Betty's Creek, etc., and you go out and ask everybody you run into what weather signs, what specific weather signs they can remember, and you be back here in an hour and forty-five minutes.' We got something like 110 weather signs in an hour and forty-five minutes that we never even suspected existed, and those appeared on three or four pages inside *The Foxfire Book*.

Another kind of interview we do is the personality type interview. We usually feature one person in each issue. In this particular issue it's a woman named Ada Kelly, who's a grandmother that lives in our section. The kids who are responsible for the Ada Kelly article go back and interview her at least three or four times . . . The kids will organize all that together and present it. Meanwhile, if the folklorist wants to see all of the original material, we have all the original transcripts plus all the original tapes . . .

With an ongoing organization, it becomes possible for each successive year of pupils to help not only each other, but following years. They are encouraged to choose their own subjects. But very often they collect much more than they need for their own purposes:

He may have gone out for ghost stories and come back with fifteen hunting stories also. When a kid gets the transcriptions all made and you have a carbon, he cuts out of the carbon the material that he needs for his article, his ghost stories . . . but somebody else in the class may be collecting hunting stories . . . He puts those together and says, 'Here, you can add these to your article.' The guy that's collecting hunting stories may do the same for the ghost-story article; in other words, they all plug in with each other. We get twenty or thirty articles usually going at the same time. If a kid comes back with a chunk of information that is not being dealt with at that time, we have what we call an 'Articles-in-Preparation File' . . . Then next year, if a kid wants to pick up that subject and carry it on . . . he can pick up that interview and go on with it. So all the material gets utilized; all of it gets used in one way or another.

Eliot Wigginton's *Foxfire* project is undoubtedly a remarkable enterprise, an example which needs to be thought about very closely. There can be no doubt that for many pupils involvement in it proves a transforming experience. As one puts it:

I've learned through *Foxfire* . . . to express myself and communicate. Then by actually teaching a younger kid how to do something I've learned to appreciate the value of teaching and become excited when I see the kid's eyes light up . . . Then more significant than that, I've learned to appreciate the value of people working together, people being dependent on each other . . . It's made a difference in my life.[2]

So much for examples of projects which have worked. But for the teacher who is considering just starting, this very success may seem daunting. What about the normal problems which can be expected in an oral history project?

First, there are those of organization. Oral history work is essentially a small group activity, and difficult to organize in large classes. There can be no doubt that projects benefit greatly from situations in which teaching is organized on a team basis, or the curriculum favours interdisciplinary work—and it is perhaps an argument for them that they point to fundamental issues in school organization. But many problems can be overcome by advance preparation—for example, by making pre-interview contact with old people, groups of children can be sent out interviewing while others are discussing material already collected in the class. For this and other purposes it is vital to draw as fully as possible on outside resources, starting with the children's own parents and the Parent–Teacher Association, and using the community networks provided by the local paper and local radio, clubs, firms, and libraries. If the school is in an area of high unemployment there will also be many skilled middle-aged people who could help the project.

Secondly, there is the question of equipment. Oral history work does not depend on the existence of recorders, although it can be much more fully developed with them. If the school's policies towards equipment and its use are not generous, you will again need to rely on the support you can win from outside, beginning with parents. There should certainly be enough homes which can lend you the tape recorders to make a start.

Thirdly, subjects have to be well chosen. They must interest each particular group of children. For younger age groups, family history is particularly suitable. It helps a child-centred approach, drawing on the child's own access to family memories and documents, and at the same time encourages parents or grandparents to participate in the work of the school. A whole variety of themes can start from developing family trees with different kinds of information. With older groups there are many more choices: homes and houses, food and clothes, work including domestic work, family life, games or leisure—and any of these can be made comparative with memories from other countries. A project can be about a local event. But perhaps one idea may be given a special mention at this point: a project which focuses on the story of a particular street. It is not always easy to see any boundaries or shape to local or community history in a big city. But a single street can offer a microcosm of some aspect of its history: of changes in working-class community life, or shops and trading, or of successive patterns of immigration. With the help of a local newspaper, in some cases it may even prove possible to trace some representative of most of the families who lived there forty or sixty years ago. It will certainly provide the teacher with a physical basis for a class project which could combine photography, the collection of records and topographical data, searches of archives and newspaper files, as well as interviewing.

Next, the children must learn some of the personal skills needed in the interview, which is not always easy. They can practise by interviewing the teachers themselves, or members of their own family. Or old people can be asked up to the school, although this normally proves much more successful in an informal small group, rather than in front of a whole class. Indeed there is a danger of leading children through such a demonstration to think of old people as historical objects, rather than valued for themselves. Children can also be asked to write their own autobiographies, partly from family documents, and be interviewed by another child—making sure that this does not lead into too deep a probing of personal situations, which could prove a damaging experience. They must also be taught how to formulate different kinds of questions. This can be helped by criticizing other interviews, like the teacher's own. And it is particularly important that the children's first interview recordings should be listened to and commented on, to give support as well as advice.

Special sensitivity is needed for projects in racially mixed schools. Clearly, both Black and White children need to learn about their own and each other's historical cultures, and there has been much recent curriculum with this objective, some of it including mutual interviewing. But for historical work it must be remembered that many immigrant children have no grandparents; and their families speak either altogether different languages, or in accents which English children find hard to understand. Getting prior support from the community will be more than usually important, and the mechanics of interviewing need to be thought through especially carefully. For example, while old White people can be interviewed by Black children in clubs, racial tension and apprehensiveness may make it very difficult to arrange for this in their homes. There is also a special need in this situation for an integrating theme—such as the relationship between industrialization and migration. It is undoubtedly an area in which more experimentation is needed.

What can be done with the fruits of oral history projects afterwards? As we shall see with community projects, they can be combined with photographs, letters and documents, clothes, tools, and other objects to mount an effective exhibition, either in the school or a local centre. They can be brought together into sound and slide lectures, or published as local booklets or newspaper articles. Afterwards, recordings can be deposited in the public library or museum. Wherever possible, a continuing, active relationship with the local community should be built up through returning to it the material collected in some form.

Finally, we need to beware against too easy success. Oral history projects can only be carried out successfully by skilled teachers in carefully considered contexts. Careless exercises could lead to considerable offence being caused, for example, by the tactless release of damaging (or libellous) information. And the advantages of the approach are largely destroyed if over-prepared, centrally produced materials are imposed. Hence the final product should not aim for a technical standard beyond the reach of the children. It is essential for them to be involved at every stage in the process, and to recognize their own contribution at the end. If oral evidence becomes just another source material in teaching, its hold on the imagination will be lost.

Less need be said at this point on projects in higher education, in colleges and universities, where there has again been much successful work in oral history. Increasingly this takes the form of individual research for undergraduate projects or postgraduate theses. The practical difficulties will depend partly on the topic chosen: in particular, if the informants needed are either hard to trace or very busy public persons; or if it takes you into strange and distant places. Thus launching research in a Third World country requires special language preparation, visas, inoculations, medical kit, knowing how to fumigate your house, how to choose—or sack—your interpreter, whether or not to pay your informants, and so on. But for most the problems will be those typical of oral history work, and also of writing projects and theses, in general. For group work too the basic approach is the same. As a starting point, for example, I have found that constructing family trees and talking about them, or interviewing in pairs and reporting back, will immediately involve and draw out almost any student group.

There are, however, two particular difficulties at this level. In principle it is possible to go beyond learning together to carry out a joint research project. With pooled effort, this can go further faster. But group projects are much more difficult if an assessment on an individual basis for each element in the course has to be returned. And even individual projects become very difficult when, because of an inflexible system which only admits closed written examinations, they cannot make a contribution to the final degree result. But raising this issue, and the need to give space for creativity within the system, is in itself a merit of project work.

A more special danger is in allowing the teaching of the method to shift too far from the practical to the abstract principle. Such academicism is largely responsible for the recurrent unpopularity of 'methodology' courses in the social sciences. Discussion of theory needs to be interwoven with practical experience, and it also needs to be directed towards specific historical issues. At the University of Essex, where the MA in Social History includes a course in the interview method, we have taught through a workshop approach. Students are plunged as quickly as possible into interviewing each other, and then seeking out an informant to interview on a topic of their own choice. These interviews are played back and discussed in the group. They invariably raise questions of the accuracy of memory, suppression, interviewing technique, and the experience of being interviewed. They also bring out examples of the kinds of historical material which can be collected, and the complexity of attitudes which it reveals. These practical sessions are interwoven with others discussing the principles of oral history from reading, so that the two approaches cross-fertilize each other.

The students then go on to mini-projects of their own, conducting a series of interviews and writing up an evaluation of their method and findings, including a model list of questions which could be used in a more extended project involving more than one interviewer. In many cases these small exploratory projects have proved the starting points for full graduate research dissertations.

The choice of the topic is clearly critical. We have found it possible for a single student, undergraduate or graduate, working alone on a vacation project, to make a real contribution to historical knowledge through new field-work. It is best (although not essential and for some subjects intrinsically impossible) if interviews can be combined with research in archives or on local newspapers. It is also important to choose a subject which is relevant to wider historical issues as well as a sufficiently defined and *localized* theme. It will fail if potential informants are too scattered to be traced relatively quickly. Some examples of subjects which have proved creative, but manageable, are studies of various East Anglian village communities; neighbourhood and class relations in a Nottingham slum and in fishing ports; migrant hop-pickers, the recruitment of schoolteachers, and the Essex military tailoring industry; young Bengali women's migration to Britain in the 1970s, and London's Italian community; domestic economy among farm labouring families, and the spreading use

of birth control among various social groups; Colchester engin-
eering workers in the First World War, and the experience of the
General Strike of 1926 in the town.[3]

The possibilities, however, are limitless. And the gains are
equally clear: the personal fulfilment, co-operative spirit, and
deeper understanding of history itself which can result—and
beyond that, the breaking down of the isolation of academic
study from the world without.

It is no doubt here—for history in the community—that the oral
history project has its most radical implications. It can contribute
to many different enterprises—an adult literacy programme, an
evening class in history, a local history society or a community
group project, a Manpower Services Commission scheme for
retraining the young unemployed, a reminiscence therapy group
for old people in a home or a hospital ward, a museum exhibition, or a radio programme. For each its essential merits will be to
encourage co-operation, on an unusually equal footing, in the
discovery of a kind of history which means something to ordinary
people. Of course these are tendencies which have to be nurtured;
and they can create problems.

The first issue concerns the choice of topics. For many of these
purposes the best topic may seem simply one which will catch
immediate interest. By contrast the local history society's notions
may be more severe, and indeed possibly too limited by the traditional conventions of documentary history. But the broadening
perspectives of social history have meant that, with imagination, a topic can certainly be found which satisfies both. Thus
Raphael Samuel has argued for a re-mapping of local community
history

in which people are as prominent as places, and the two are more closely
intertwined. He or she can then explore the moral topography of a
village or town with the same precision which predecessors have given to
the Ordnance Survey, following the ridge and furrow of the social environment as well as the parish boundaries, travelling the dark corridors
and half-hidden passageways as well as the by-law street. Reconstructing
a child's itinerary seventy years ago the historian will stumble on the
invisible boundaries which separated the rough end of a street from the

respectable, the front houses from the back, the boys' space from the girls'. Following the grid of the pavement you will come upon one space that was used for 'tramcars', another for hopscotch, a third for Jump Jimmy Knacker or wall games. 'Monkey racks' appear on the High Street, where young people went courting on their Sunday promenades, while the cul-de-sac becomes a place where woodchoppers had their sheds and costers dressed their barrows . . . (And similarly in) particular woods or fields . . . here mushrooms could be found or rabbits trapped; there potatoes were dug or horses illegally grazed, or long summer days were spent at haymaking or harvest. . . .

Or again, instead of taking a locality itself as the subject, the historian might choose instead as the starting point some element of life within it, limited in both time and place, but used as a window on the world . . . It would be good to see this attempted for nineteenth-century London. A study of Sunday trading in Bethnal Green, including the war waged upon it by the open-air preachers; of cabinet-making in South Hackney, or of Hoxton burglars . . . would take one closer to the heart of East End life than yet another précis of Hector Gavin's *Sanitary Ramblings* . . . Courting and marriage in Shepherd's Bush, domestic life in Acton, or Roman Catholicism among the laundrywomen and gasworkers of Kensal Green might tell one more about the growth of suburbs than logging the increase of streets . . . The study of social structure, too, might be made more intimate and realistic if the approach were more oblique, and focused on activity and relationships. A study of childhood in Chelsea (of whom you could or couldn't play with, of where you were allowed to go), masculinity in Mitcham, the journey to work in Putney, or of local politics in Finsbury, would tell us a great deal (more) about the way class differences were manipulated and perceived, and social allegiances expressed in practice . . . than a more flat-footed approach taking the Registrar-General's fivefold divisions as markers.[4]

It is this approach of seeking for 'a window on the world' which initially allowed Raphael Samuel to catch the imagination of trade unionists on adult education courses at Ruskin College, Oxford, by getting them to explore the histories of their own occupations, and subsequently to stimulate the History Workshop movement which spawned local groups through the districts of London and the provincial cities. The movement's half-yearly journal, *History Workshop*, should provide sufficient reassurance to those pessimistic enough to fear that enthusiasm must prove incompatible with scholarly standards.

Its activities have nevertheless presented a challenge to professionalism as such, 'dedicated to making history a more democratic activity', and attacking a situation in which ' "serious history" has become a subject reserved for the specialist . . . Only academics can be historians, and they have their own territorial rights and pecking orders. The great bulk of historical writing is never intended to be read outside the ranks of the profession.'[5] A similar opposition to this view underlies the activities of many oral history groups, such as the collection of songs and interviews from north Italian factory workers by the Istituto Ernesto di Martiro in Milan, the joint work of the Brighton Trades Council and Sussex Labour History Society in the Queen Spark Books series, or other co-operative life story publishing groups such as Peckham People's History, Bristol Broadsides, and the People's Autobiography of Hackney from East London. There is a parallel too in the neighbourhood committees which were set up in fifteen districts of Boston to produce, from library research, locating photographs, and collecting memories from all kinds of people, a bicentenary series of history booklets. There can be no doubt that these booklets, which were distributed free to residents in the Boston Bicentenary series or sold several thousand copies in the case of Hackney, have brought local history to an exceptionally wide audience. But the spirit in which they are produced by co-operative work can be equally striking. In the Hackney group there was an insistence that anyone can record anyone else, and all should contribute to the process of presentation. The purpose was as much to give people confidence in themselves, and their own memories and interpretations of the past, as to produce a form of history. In this context the professional, confident in self-expression and backed by the authority of higher degrees, can become a positive menace, tearing at the roots of the project. Of course a complete absence of the wider historical perspectives of an experienced historian can be equally damaging to a group's work. It will lead to the creation of one-dimensional historical myths rather than to a deeper social understanding. What is needed is a dynamic relationship, with interpretation developing through mutual discussion.

The success of the local group project will thus depend partly upon how it makes use of the differing talents which each will bring to the work: their own life memories, ability as technicians, knack with organizing, or skill in getting others to talk, will be as important as a reservoir of historical information. In some projects roles can be best divided up within a normal committee structure; in others a professional will be the informal leader of an egalitarian group; in yet others, such as Manpower Services Commission projects for the young unemployed, the team will be formally contracted employees, working under a supervisor.

This last type of project has to be operated within a special framework, with clearly laid down objectives and bureaucratic procedures. In order to secure funding it must offer not only a list of useful end-products for the local community, but training for young people: in interviewing, audio-typing, secretarial work and accounting, word processing, handling recording equipment and tapes, as so on. An oral history project can be readily designed to achieve this, but the double objective, and also the strict rules which define age, qualifications, recruitment, and salaries, will shape the staffing structure. There must be a secretary to deal with the contracts, time sheets, cash books, accounts, and other obligatory paperwork, and a high ratio of part-timers will be needed to allow reasonable rates of pay under the rules. The project's key appointment will be its supervisor, who should be an unemployed professional with good qualifications and experience; but 'you can afford to be more flexible and experimental when recruiting other project workers. People with little or no formal qualifications often possess a real flair for this type of work, especially the interviewing and recording.'[6] A more fundamental problem, however, is that the project will only be funded for one year, and even if renewed almost all the staff will have to be replaced by new trainees. Thus the team has often just got to the point when it is working well, only to be cut off. It is hardly surprising that the results of Manpower Services projects have been uneven. They have nevertheless been probably the biggest single source for the expansion of local oral history archival work in recent years. And where a series of projects has allowed the cumulation of expertise, as in the combined neighbourhood collecting of oral memories and family photographs by Manchester Studies in the 1970s, or on immigrant groups by Bradford Heritage in the 1980s, they can bring high rewards.

The level of equipment will also determine what can be achieved. As we shall see, both interviewing and archiving demand good equipment to get the best results. Certainly a life story group can run very successfully on no more than pencil and paper. But if you want to record for local radio or set up an archive, the equipment you really need is expensive, and this is best recognized at the start. Manpower Services schemes are severely under-capitalized; and without a supportive sponsor, such as a local library, museum, or educational resources centre, to provide accommodation and much of the equipment, there will be scarcely enough money for the phone and post. Yet equipment is all the more necessary if the project is to provide training. Before starting any project it is essential to know that you can either borrow or buy workspace, desks, chairs, filing cabinets, at least two good tape recorders, and a typewriter or word processor.

The next choice is whose memories to record. This is of course critical for any oral history project: and the underlying principles remain the same. First, there is little point in recording people whose memories are confused or impaired, or who are too withdrawn to talk about them. Secondly, what matters is the *direct* personal experience that somebody has, rather than their formal position. This is a particular trap for local historical societies or public libraries. It can mean that the people chosen to record are those very local dignitaries, such as mayors and council officials, who have the most need for caution and thus the least to give. It is 'almost axiomatic', as Beatrice Webb very correctly observed,

that the mind of the subordinate in any organization will yield richer deposits of fact than the mind of the principal. This is not merely because the subordinate is usually less on his guard . . . The working foreman, managing clerk, or minor official is in continuous and intimate contact with the day-to-day activities of the organization; he is more aware of the heterogeneity and changing character of the facts, and less likely to serve up dead generalization, in which all the living detail becomes a blurred mass, or is stereotyped into rigidly confined and perhaps obsolete categories.[7]

Thirdly, it is necessary to be constantly aware of the social balance of the accounts which are being collected. Thus there is always a tendency for projects to record more men than women. This is partly because women tend to be more diffident, and less often believe that their own memories might be of interest. It is also because men are much more often recommended as informants by others. Even when this is recognized as a problem, it may prove difficult to solve. For example, if the subject is a local industry, it will be easy enough to find men who worked in it; indeed they may still meet together as old workmates at a pub or a club. But their wives, or women workers in the same industry, although equally vital to its functioning, will be much harder to trace, because they will not normally be locally known by their occupation, and their social networks will be those of the neighbourhood rather than the workplace. Similarly, there is an equally strong tendency for a community project to record its central social stratum—normally the respectable working class and the lower middle class—at the expense of both top and bottom. There are difficulties in tracing the retired works director to Cheltenham Spa. And again and again, the very poorest, the 'rough' elements which were a vital part of the community, prove equally elusive. They are not suggested as informants because the more 'respectable' old people either positively disapprove of what they would say, or simply regard them as too pathetic or unintelligent to have any worthwhile memories at all. Yet they are often precisely those whose different vi— richly expressed in dialect stories, can provide the most valuable recording of all. And it is the juxtaposition of live experience, from all levels of society, which makes the most telling and thought-provoking local history.

Finding a sufficient range of informants is thus a key task. A self-selected group, responding to a public notice or a local newspaper or radio appeal, can certainly provide the best start for some projects, but it will rarely be representative enough. People can be located in many other ways: through personal contacts; at old people's workshops or clubs; through trade unions or political parties; appeals in the local paper, a shop window or on the radio; through welfare workers or doctors, churches or visiting organizations; even by chance encounters in a shop or a pub or a park. It is always much easier when you can approach them with a personal recommendation from somebody else. Although there will be refusals, which can be disheartening, provided that you keep a clear idea of who you are searching for, this part of the project depends above all upon persistence. But it will be worth persisting.

Lastly, what about the outcome? For the future, your tapes and transcripts must be deposited, along with photographs, documents and other material you have collected, as a resource for future public use: and the best place for this is likely to be your public library. More immediately, you can draw on them to produce educational 'jackdaw' packs for use in local schools, including cassettes of interview extracts; to stage small travelling exhibitions, again combining sound with photographs and text; and to make a tape and slide show for use with old people to stimulate their own memories of the community's past. The tapes can also be used to make programmes for local radio—for these, a collage of extracts with a minimal linking by a narrator will often be the most effective form; or you may be able to find a drama group with whose help you can develop some of your material into a stage production. And you can publish your material in printed form: as newspaper features, as local booklets, or—like Island History in London's Isle of Dogs, or the East Bowling History Workshop in Bradford—as annual calendars of old photographs.

With this last form, only the caption may betray its origin in oral recording work; and this may be the case in other outcomes. Thus many museums have used oral history projects for reconstructing, correcting, and interpreting material displays, from the Smithsonian's Black tenant house in Washington, or the Schneider steel furnaces which once lit up the company town of Le Creusot in France, to the 1920s Durham dentist's surgery at Beamish; but using the oral evidence from the project in the display itself, as in the fishing family's kitchen at Lancaster Maritime Museum, is still strangely rare, though equally effective. Thus at the Gloddfa Ganol slate mine in Blaenau Ffestiniog, North Wales, you can watch and hear miners themselves in a tape and slide show in the old stables before entering the vast caverns cut within the mountain. Elsewhere visitors may be lent cassettes to carry round the building themselves. The extract is best when very brief, less than a minute. A labourer can speak about ploughing, or a weaver explain a loom. In the Imperial War Museum you can hear the sound of marching and guns as you look at exhibits from the trenches, or enter a recruiting booth where an old soldier's voice recounts what it was like to be signed on.

Even a temporary initiative may have a remarkable impact. Both Coventry and Southampton museums organized exhibitions from their projects which included tea dances to jazz bands, at which old couples spun the floor in rejuvenated delight to the sound of long-lost tunes; and these occasions also proved catalysts for the reunion of old workmates, and even of two sisters who had not seen each other for forty years. In other projects, old people's reminiscence groups have contacted local schools as well as vice versa. In Harlow New Town, Jewish teenagers ran an old people's club inside their own club; while in the Cambridgeshire village of Burwell a project for a documentary film including recorded memories led to a 'drop-in club', again within the youth club, for old people collecting their pensions from the post office on Thursday afternoons. They chatted with some of the youth group over tea, leading on to recordings, articles published in the community magazine, and so more offers of information.

A small number of community projects are much more ambitious—or succeed beyond expectations. Thus the New York Chinatown History Project aims to help build a democratic community structure precisely by a highly politically conscious but sensitive interpretation of older and newer immigrants, community bosses and sweated laundry workers, to each other. In Sweden a project in the condemned slum neighbourhood of Nöden in the city of Lund in the 1960s so revived local community feeling that Nöden was permanently saved from the motorway project which would have destroyed it; and twenty years later the 'Dig Where You Stand' factory history tent tours the country stirring the solidarity of old industrial communities in a similar way. And a British project, by evoking very different memories, succeeded in turning the local radio into a two-way channel. For this Dennis Stuart of Keele University's Adult Education Department conceived a project in local Methodist history, working with Arthur Wood of Radio Stoke-on-Trent. A set of eight chapel study groups was formed, each examining their own records and carrying out interviews, and this material was brought together, linked by narration and live recordings of singing and preaching, in a series of broadcast programmes. The programmes, and an exhibition linked with them, stimulated more recording work—and also a new programme series, consisting of weekly fifteen-minute programmes, each composed of a patchwork of voices with no narrative, mostly on aspects of social life before the First World War, which proved very popular, and lasted altogether eighty weeks. This again stimulated very active local participation, with people sending in comments, offering to be interviewed, and bringing in essays about their own memories.

We can end with a still more striking instance of local participation, which has brought about the creation of a whole new museum in Italy. This is not an oral history project as such, so much as a community history project in which the role played by oral memory was of vital importance. It also provides a very striking instance of the co-operation possible between workers and university scholars, taking place in the context of a wider Italian political movement for the re-evaluation of working-class culture. As Alessandro Triulzi, in describing the project, argues, the Museum of Peasant Civilization—the Museo della Civilta' Contadina di S. Marino in Bentivoglio—must be understood not just as a collection of objects from the past, but as 'the workers' own answer to the cultural appropriation they have been subjected to by the dominant classes', and a step towards the 're-appropriation of values and contributions that have long been ignored or trivialized and distorted by the state official culture'.

The museum stands some miles outside Bologna, in the hot, flat, rain-swept countryside of the Po valley. It is a superbly documented and socially very penetrating display, set out in the elegant bailiff's house of a former landed estate, of the life and work of the peasant sharecroppers of the surrounding country side. Opened in 1973, it was the fruit of a campaign which had lasted nine years. It began in 1964 when a former peasant, Ivano Trigari, discovered an old farming tool half-buried outside a friend's house, locally known as a *stadura*:

The *stadura* is a round iron bar, fifty or sixty centimetres long, which was used in old Bolognese ox-carts both as a brake and as an ornament. The top of the bar was usually embellished with a cross or other decorations, and it had one or more iron rings which gave each cart when moving its own characteristic sound.

Trigari cleaned and polished this *stadura*, and put it on show in the window of the agricultural co-operative where he worked in the little town of Castelmaggiore. The result was astonishing: a 'fever for *stadure*' seized the town, a competition to produce the best specimen, schoolchildren bringing in examples sent by their fathers wrapped in paper, news of discoveries coming from every corner. Within a few days some twenty had been amassed, the most beautiful on display in the front of the shop, the rest piled in a corner. The collection became the talk of the town, drawing crowds of old peasants from the bar and the workers' club, the Casa del Popolo. As they stood there looking and commenting, Trigari listened to their memories, questioning them for details he did not know. He realized, as he puts it, that these oral memories could provide 'a general fresco of an epoch which had disappeared already, or was soon to disappear'. The comments varied:

Some cursed the tools which reminded them how hard they had worked in the past; others were excited, reminded of their youth. They said that times were better now, and started exchanging memories of the past, of times when they had to rise at two in the morning to go ploughing; of how they had to take their ox-carts to the rice fields to collect forage and the rice-straw which was used then as litter for the animals . . . And again of when they used to take all the hemp to their landlord's mansion or how they carried the huge grape baskets on their heads; or of the great bundles of firewood which were carried to the baker; or finally, of when, with the best cart and oxen, the *stadura* all shining, the bridegroom went to the bride's house to take her dowry.

Before long, other tools were coming in too: old looms, hemp tools, yokes, hoes, ploughs, and so on. Out of this grew the idea of a systematic local collection of old working implements. It was taken up enthusiastically by the peasants, who hunted through their houses and in disused depots, and persuaded their friends to do the same. A *festa della stadura* was started to win support, which has become an annual occasion. An association was set up, the *Gruppo della stadura*, which organized a travelling exhibition, an ancient cart drawn by a tractor with samples of tools, putting them on show in nearby villages at fair and carnival times, or at the celebrations of local saints' days, and appealing for support. 'Peasants would listen, and often contributed on the spot tools, money, advice, and suggestions where further material could be found.' By the time that, after years of searching, the association won the present home for its museum from the provincial administration of Bologna, almost 4,000 objects had been assembled. 'Based on an unfailing faith and pride in their own sense of history', the movement had developed into 'a collective effort that involved almost everybody in the community'.

A key factor in winning official support had been the backing of a group of university students and researchers gathered around Carlo Poni, Professor of Economic History at Bologna. With the first stage of the project brought to success with the opening of the museum, the close contact between historians in the university and the community is intended to continue. The museum is building up a substantial archive of labour contracts, estate papers, the records of peasant organizations, and photographs. It acts as a centre for seminars and for research in agrarian history. At the same time it draws in thousands of local visitors, especially schoolchildren, who are encouraged to write their own term papers using the museum material. It is also providing training for others who wish to start similar ventures elsewhere: some fifteen new agrarian museums are being created in Emilia alone, and the idea is spreading to other provinces. Perhaps most interesting, for the oral historian, is the encouragement of local historical memory. Schoolchildren collect interviews from their own villages as well as using the museum's records. The interview method has also been developed by university students. And above all, peasant-historians within the community have found a renewed confidence. One instance will suffice to conclude: that of Giuseppe Barbieri of S. Giovanni in Persiceto.

Aged 78, Guiseppe Barbieri, now a retired peasant, has a poor scholastic record (he didn't get beyond third grade) but a considerable record as a local historian. His first work, a 300-page manuscript titled 'My memories in war and peace. Some family remembrances', was written in 1936. In it the then 39-year-old peasant described at length this war experience, the peasants' working conditions in pre-WW1 Emilia, the agrarian struggle of 1919–20, and peasant reactions to national events, like the rise of Fascism, or to local tragedies, like the earthquake of 1929. Written in ungrammatical Italian (almost a foreign language to him since, like most peasants in the area, he speaks the local dialect both at home and work), his exercise book manuscript lay idle in his house until 1975, when news of the peasant museum spread in the region . . . Guiseppe Barbieri decided to take up pen again . . .

Barbieri's new book was to be on rural traditional structure and daily labour. It was to offer, in Triulzi's words, 'a working man's answer to the scholars' false dichotomy between the little history of daily life and labour and the Great History of official textbooks'. And Guiseppe Barbieri certainly entered his task with enthusiasm; for he hoped, in his own words, 'to continue it quickly, while my memory is still good, since I have passed my 77th year already and I feel proud to express our past'.[8]

The spirit in which to undertake an oral history project could hardly be better put.

# 7

# The Interview

To interview successfully requires skill. But there are many different styles of interviewing, ranging from the friendly, informal, conversational approach to the more formal, controlled style of questioning, and good interviewers eventually develop a variation of the method which, for them, brings the best results, and suits their personality. There are some essential qualities which the successful interviewer must possess: an interest and respect for people as individuals, and flexibility in response to them; an ability to show understanding and sympathy for their point of view; and, above all, a willingness to sit quietly and listen. People who cannot stop talking themselves, or resist the temptation to contradict or push an informant with their own ideas, will take away information which is either useless or positively misleading. But most people can learn to interview well.

The first point is the preparation, through reading and in other ways, of background information. The importance of this varies a good deal. The best way of starting off some pieces of work may be through exploratory interviews, mapping out a field and picking up ideas and information. With the help of these a problem may be defined, and some of the resources for solving it located. The 'general gathering interview' at the beginning of a local project, like the 'pilot interview' of a big survey, can be a very useful stage. And of course there is no point in having any interview at all unless the informant is, in some sense, better informed than oneself. The interviewer comes to learn, and indeed often gets people to talk in just this spirit. For example, Roy Hay has found in his research with the Clydeside shipbuilders that quite often, 'one's own ignorance can be turned to good use. On many occasions older workers have greeted my naïve questions with amused tolerance and told me, "Naw, naw laddie it wasn't like that at all," followed by a graphic description of the real situation.'[1]

It is nevertheless generally true that the more one knows, the more likely one is to elicit significant historical information from an interview. For example, if the basic narrative of a political decision, or a strike, has been established from newspapers, it will be possible to place the informant's own part within the events exactly, to identify where he or she may have special direct experience and observations; which recollections are second hand, and to spot elisions of memory between similar events at different times—like the two 1910 General Elections, or the strikes of 1922 and 1926. This background information may itself have been built up very fully through earlier interviews, as with the systematic reconstruction of Jewish persecution and resistance during the Second World War, or of the local partisan movements in Italy, where the point of a testimony may be to corroborate and fill out in precise detail the hourly events of a day in 1944 when a man's family was destroyed.

A similar command of detail can be built up for a life story interview, when the subject is a public personality, or a writer, or possesses sufficient personal papers. Although some of the material—like the subject's own writings—will be available before the start of interviewing, more may be produced as a result of the first interviews, leading to correspondence, the discovery of further documentation, and eventually more interviews at another level of questioning. Of course not all prominent informants are willing to be subjected to a step-by-step research process. Thomas Reeves found that interviewing American liberal intellectuals required exceptionally painstaking and thorough preparation. They were often too busy to grant more than brief interviews, so that 'specific, highly-informed questions' were essential. Worse still, should he 'appear hesitant, or seem to be fishing blindly, the relationship between the participants in an interview can be quickly destroyed. Liberal intellectuals seem especially interested in testing your credentials to be an oral historian by probing your knowledge of the subject under discussion. I have often felt, particularly at the beginning of an interview session, that they were interviewing me . . . These sort of queries are ploys in status games.'[2]

Such demanding informants are rare. Nevertheless, even with

a more general historical study of a community or an industry, it is important to pick up a knowledge of local practices and terminology as quickly as possible. John Marshall, for example, points out how misleading the question 'At what age did you leave school?' could be in the Lancashire cotton towns. A former millgirl would answer, at 14; and it was only because he knew that the majority had been half-timers in the loom-sheds long before leaving school—a fact which they took for granted—that he then went on to ask, 'When did you begin work?'[3] Many oral historians have found that a basic knowledge for work terms is useful, as a key to establishing mutual respect and trust. And Beatrice Webb, decades earlier, made the same point with a characteristic sharpness:

To cross-examine a factory inspector without understanding the distinction between a factory and a workshop . . . is an impertinence. Especially important is a familiarity with technical terms and a correct use of them. To start interviewing any specialist without this equipment will not only be a waste of time, but may lead to more or less courteous dismissal, after a few general remarks and some trite opinions . . . For technical terms . . . are so many levers to lift into consciousness and expression the more abstruse and out-of-the-way facts or series of facts; and it is exactly these more hidden events that are needed to complete descriptive analysis and to verify hypotheses.[4]

Nor is this true only of the specialist. It is an equal 'impertinence' to subject numbers of working people in a community or an industry to questions, without first ensuring, as far as possible, that they are historically relevant and correctly phrased for the local context.

A broader study of social change, depending on a relatively wide spectrum of informants, also demands particularly careful preparation of the form of questions before interviewing. Asking questions in the best way is clearly important in any interview. This is, however, an issue which can raise strong feelings among oral historians. A contrast may be made between so-called box-ticking 'questionnaires' whose rigidly structured logical patterns so inhibit the memory that the 'respondent'—again the choice of term itself suggestive—is reduced to monosyllabic or very short answers; and, at the other extreme, not so much an 'interview' at all, but a free 'conversation' in which the 'person', 'tradition-bearer', 'witness', or 'narrator' is 'invited to talk' on a matter of mutual interest.[5] The truth is that it takes great deftness, and a well-chosen informant, to be able, like George Ewart Evans, to get outstanding material while remaining 'relaxed, unhurried', and giving the informant 'plenty of time to move about . . .' 'Let the interview run. I never attempt to dominate it. The least one can do is to guide it and I try to ask as few questions as I can . . . Plenty of time and plenty of tape and few questions.'[6] Those few questions are based on long experience, combined with a clear idea obtained in advance of what each particular informant may tell.

The strongest argument for a completely free-flowing interview is when its main purpose is not to seek information or evidence of value in itself, but to make a 'subjective' record of how one man or woman looks back on their life as a whole, or part of it. Just how they speak about it, how they order it, what they emphasize, what they miss out, the words they choose, are important in understanding any interview; but for this purpose they become the essential text which will need to be examined. Thus the less their testimony is shaped by the interviewer's questions, the better. However, the completely free interview cannot exist. In order to start at all, a social context must be set up, the purpose explained, and at least an initial question asked; and all these, along with unspoken assumptions, create expectations which shape what follows. Experiments with this approach have generally proved disappointing: 'it tended to result in a brief, even terse account', Janet Askham found, simply because 'they did not know what I was interested in'. Stories flowed much more freely once she started to ask questions.[7] Some compromise is necessary, even for this objective.

At the other extreme, the classic survey's search for 'objective' evidence points towards a mirror of incomprehension. The purpose of an interview should be to reveal the sources of bias, which are basic to social understanding, rather than to pretend they can be nullified by a dehumanized interviewer 'without a face to give off feelings'.[8] No oral historians have in fact argued to my knowledge for the rigid inflexible questionnaire style of interview.

It is really the needs which follow from a particular type of research which make advance planning of questioning essential—for example, in any project where interviewing work is shared in a team, or where paid interviewers are used; or wherever material is to be used for systematic comparisons. The merits and drawbacks of the 'two schools of interviewing' are nicely summarized in a more qualified contrast by Roy Hay:

Firstly there is the 'objective/comparative' approach usually based on a questionnaire, or at least a very highly structured interview in which the interviewer keeps control and asks a series of common questions to all respondents. The aim here is to produce material which transcends the individual respondent and can be used for comparative purposes . . . In the hands of flexible, sensitive interviewers, who are prepared to abandon the script when necessary, this approach can generate very useful material indeed, but it can be deadly. Promising lines of inquiry are too easily choked off and, worse still, people are forced into the predetermined framework of the interviewers and so large relevant areas of experience are never examined at all.

At the other extreme is the free flowing dialogue between interviewer and respondent, with no set pattern, in which conversation is followed wherever it leads. This method occasionally turns up the very unexpected and leads to completely new lines of inquiry but it can very easily degenerate into little more than anecdotal gossip. It can produce miles and miles of useless tape and impossible problems of selection and transcription.[9]

Beyond this, there is also the effect of the personalities involved in each particular interview. Some interviewers are naturally more chatty than others, and can draw out an informant this way (although this is relatively unusual, and a more common effect of chattiness can be to shut people up). And informants vary from the very talkative, who need few questions, just steering, or now and then a very specific question to clarify some point which is unclear; to the relatively laconic, who with encouragement, broad open-ended questions, and supplementary prompts, can reveal much richer memories than at first seemed possible.

There are a few basic principles in the phrasing of questions which apply in any interview. Questions should always be as simple and as straightforward as possible, in familiar language.

Never ask complex, double-barrelled questions—only half will usually be answered, and it usually won't be clear which half. Avoid a phrasing which points to an unclear answer: for example, ask, 'How often did you go to church?', rather than, 'Did you often go to church?' Of course occasional hesitation does not matter at all, and may even win a little sympathy from the informant. But frequent apologetic confusion is simply perplexing, and is especially to be avoided as a style of asking delicate personal questions since it only conveys your own embarrassment. A careful or indirect question, previously worked out and confidently put, is much better. It shows you know what you're doing, so the atmosphere is more likely to stay relaxed.

You will need a different kind of phrasing to establish specific facts and to get description or comment. The latter demands an 'open-ended' type of question, like 'Tell me all about . . .', 'What did you think/feel about that?', or 'Can you describe that to me?' Other cue words for this sort of question are 'explain', 'expand on', 'discuss', or 'compare'. If it is a really important point, you can encourage at length: 'All right, so you're in—. Shut your eyes, and give me a running commentary—what you see, hear . . .' A physical description can also be suggested as a lead into an evaluation of a person's character. Right through the interview whenever you get a bald fact which you think might be usefully elaborated, you can throw in an inviting interjection— 'That sounds interesting'; or more directly, 'How?', 'Why not?', 'Who was that?' The informant may then take up the cue. If, after some comment, you want more, you can be more emphatic ('That's very interesting'), or mildly challenging ('But some people say that . . .'), or try a fuller supplementary question. In most interviews, it is very important to use both kinds of questions. For example, you may be told, as a general comment, that 'we helped each other out', 'we were all one big family in the street', but if you ask a specific question such as who outside the family helped when the mother was ill, it may become clear that neighbourly aid was less a practice than an ideal. Getting behind stereotyped or non-committal generalizations to detailed memories is one of the basic skills, and opportunities, of oral history work.

Leading questions must normally be avoided. If you indicate your own views, especially early in an interview, you are more likely to get an answer which the informant thinks you would like to hear, and will therefore be unreliable or misleading as evidence. There are some exceptions to this. If you know somebody has very strong views, especially from a minority standpoint, it may be essential to show a basic sympathy with them to get started at all. Also, in order to allow the possibility of some responses which would be conventionally disapproved by most people, it may be best to ask a loaded question: 'Can you tell me of a time when you had to punish severely . . .?', 'Were most people taking home materials from the works in those days?' or 'I hear the mayor was a very difficult man for his immediate employees to get along with'—which is much more likely to provoke a clear reaction than a bland form like, 'I know the mayor was a very generous and wise person. Did you find him so?'[10] But such questions are dangerous on most occasions, and are not normally appropriate. Most questions should be carefully phrased to avoid suggesting an answer. This can be quite an art in itself. For example, 'Did you enjoy your work?' is loaded; 'Did you like your work or not?', or 'How did you feel about your work' are neutral.

Finally, avoid asking questions which make informants think in your way rather than theirs. For example, when dealing with concepts like social class, your information is much stronger evidence if you encourage them to produce their own basic terms, and then use these in the subsequent discussion. And try to date events by fixing the time in relation to their own age, or a stage of life, such as marriage, or a particular job or house.

Even if you are going to carry out only a small oral history project of your own, it is worth thinking about the sequence of topics for the interviews and the phrasing of questions. The strategy of the interview is not the informant's responsibility, but yours. It is much easier to guide if you have a basic shape already in your mind, and questions can lead naturally from one to the other. This also makes it easier, even when you digress, to remember what you still need to know about. In addition, for most projects you will need some basic background facts from all informants (origins and occupations of mother and father, own birth, education, occupations, marriage, and so on), and you will also find a recurrent need for basic and supplementary questions on many topics. If you have already worked these out in your head, and can toss them in when needed, you can more easily concentrate on what the informant is saying, instead of trying to think of how to get in yourself.

For many purposes a list of headings jotted down for reminders on less frequent topics is enough. But for teamwork, or for a comparative project on any scale, a more fully elaborated interview schedule is desirable. An example, complete with directions for interviewers, is illustrated in an appendix of Model Questions. Provided such a schedule is used flexibly and imaginatively, it can be advantageous; for in principle, the clearer you are about what is worth asking and how best to ask it, the more you can draw from *any* kind of informant. With relatively reticent people, who say right at the start, 'It's all right as long as you use the questions', this is obvious enough: and such informants are quite common. You can then follow the lines of the schedule more or less methodically. With very talkative people the schedule should be differently used. If they have a clear idea of what they want to say, or the direction the interview should take, follow them. And wherever possible avoid interrupting a story. If you stop a story because you think it is irrelevant, you will cut off not just that one, but a whole series of subsequent offers of information which *will* be relevant. But sooner or later, they will exhaust their immediate fund of recollections, and they too will *want* you to ask questions. With this kind of informant several visits will be needed, and afterwards you can play back your recordings, checking against the schedule what has been covered and what is worth asking in later sessions. The printed form of the schedule in this case becomes particularly useful. But normally it is much better to know the questions, ask them directly at the right moment, and keep the schedule in the background. Essentially it is a map for the interviewer; it can be referred to occasionally, but it is best to have it in mind, so that the ground can be walked with confidence.

Certain other decisions need to be taken before the interview.

First, what equipment should be taken? In a minority of contexts, the best answer is none. Even note-taking, let alone tape recording, may arouse suspicion in some people. Fear of tape recorders is quite common among professionals whose work ethic emphasizes confidentiality and secrecy, like civil servants, or bank managers.[11] For different reasons it can also be found among very old people, who feel hostile to new technology; among minorities who have experienced persecution, and fear that any information on tape might get into the hands of the police or authorities, and be used against them; or in close-knit communities where gossip is feared. Some people may object to recording, but not to note-taking. Even if neither is possible, a skilled interviewer can learn to hold enough of the main information and key phrases to jot down soon afterwards, and make an interview worthwhile. Indeed, before tape recording made such a method seem by comparison impressionistic, this was the commonest sociological practice.

Most people, however, will accept a tape recorder with very little anxiety, and quickly lose any immediate awareness of it. The recorder can even help the interview. While it is on, people may be a little more likely to keep to the point, and other members of the family to stay out of the way. And quite frequently, when it is switched off, some highly significant additional facts may be given, which could have been held back if there had been no recorder at all; information which is meant to be known to the researcher as background, but in confidence (and must of course be treated in this spirit). When using a tape recorder it is important to avoid drawing attention to the machine, and diverting one's own attention, by fussing about it. If it is a new one, make sure you have read the manual which goes with it, had somebody demonstrate its workings, and practised operating it and setting it up. Check before you set out that it is functioning, and that you have not only all the parts and tapes you need, but batteries and adaptor plugs.

You can also take with you various aids to memory. An old newspaper cutting, or a local street directory can help. George Ewart Evans often takes a work tool. 'In the countryside I often take along an old serrated sickle. With that there is no need of any abstract explanation of what you are going about. He sees the object, and if you choose well he won't need any prodding to open up. We are both right into our subject from the beginning. In the same way if I was going to see an old miner, I'd take a pair of yorks or a tommy-box.'[12] Since the focus of his interviews is the work process, such a tool is an ideal starting point. If it was to be childhood in the family, a piece of clothing might be better; or for a political life story, an early pamphlet. These might also stimulate the production of old letters, diaries, cuttings, and photographs, which is particularly worth encouraging and could be the most valuable by-product of an interview.

Next, where should the interview be held? It must be a place where the informant can feel at ease. Normally the best place will be their own home. This is especially true of an interview focusing on childhood or the family. An interview in a workplace, or in a pub, will activate other areas of memory more strongly, and may also result in a shift to a less 'respectable' style of speaking. A trip round the district can also prove rewarding, and stimulate other recollections.

Nearly always, it is best to be alone with the informant. Complete privacy will encourage an atmosphere of full trust in which candour becomes much more possible. This is usually true even of an old married couple who are particularly close to each other. Of course it is not always easy to find a tactful way of seeing them apart. (It is easier if you interview both of them; and particularly if two of you arrive together at the couple's home, and then pair off in different rooms.)

The presence of another person at the interview not only inhibits candour, but subtly pressurizes towards a socially acceptable testimony. Fortunately, however, this is not all disadvantage. An old couple, or a brother and sister, will often provide corrections of information which are positively helpful. They can also stimulate each other's memory. This effect becomes still more marked when a larger group of old people get together. There will be a much stronger tendency than in private to produce generalizations about the old days, but as they argue and exchange stories among each other, some fascinating insights can emerge. Of course the stories, even more than usual, must be understood partly as art forms, conveying symbolic meanings—indeed many more of them are likely to be about *other* people. But sometimes a group, for example in a public bar, may be the only way into a hidden world of a common work-experience of sabotage or theft, or the secret devices of poachers in the countryside.

The group can also be a useful device in other situations. John Saville and a research student met with three leaders of the Manchester Unemployed Workers Movement of the 1930s, and in five hours of co-operative discussion reconstructed many of the gaps in the newspaper evidence which they had previously assembled. With more self-defended public figures, such as Canadian politicians, Peter Oliver has found cross-examination by two or even three interviewers effective, and David Edge used a triangular interview for his work on radio-astronomers. Beatrice Webb, although strongly favouring privacy for the normal interview, also developed a technique of 'wholesale interviewing' in the more relaxed atmosphere of social occasions, once with a party 'even telling fortunes from their hands, with all sorts of interesting results!' At the dinner-table or in the smoking-room she found that 'you can sometimes start several experts arguing among themselves; and in this way you will pick up more information in one hour than you will acquire during a whole day in a series of interviews'.[13]

Once the preliminary decisions have been made, you have to make contact with your chosen informant. You can write (enclosing a stamped addressed envelope), or sometimes call in person or telephone. It will always be much easier if you can say that somebody else in the informant's own social network has recommended them. You need to explain very briefly the purpose of the research. Suggest a possible time for a first visit, but always leave the informant the chance to propose another, or to refuse altogether. With a minority of informants, like politicians or professionals, it may be wise to set out your research proposal and the use you intend for the interview more fully. This will help them to decide whether to see you, and will clarify your future right to use of the material. Some may begin thinking about the topics which interest you and search out some old papers before you come.

Most people would be more likely to find too long a letter forbidding, so it is best to wait until your first meeting. Then start by explaining the subject of your project or your book, and how the informant can help you. Many people will protest that they have nothing useful to tell you, and need reassurance that their own experience is worthwhile, that it is unfamiliar for younger people whose lives have been very different, and essential for the making of real social history. Some may be genuinely surprised at your interest, and you will need to be more than usually encouraging in the early stages of the interview. Some will explicitly raise the question of confidentiality, and not want their names given. Be open about your intentions, and honour any promises you make. Most people will trust you to be discreet with what they tell you—and this trust must be respected. Do not attach their names, without their explicit consent, to damaging quotations about themselves or their neighbours.

The start of this first meeting is also normally the best time to ask whether the interview may be tape-recorded, although sometimes this can be suggested in the initial approach. Some oral historians believe in using the first meeting as a brief, exploratory visit, for preparing and getting to know an informant, without using a tape recorder. The drawback is that, even in trying to establish basic facts about the informant's background, it is difficult not to tap the essence of the memory. You can go over the same ground on a second visit, but it is likely to be presented in a much more stilted way. In my own experience, it is best to get the recorder going as quickly as you can once you get talking.

This raises another matter controversial among oral historians —the quality of recording. For a really good recording, of the quality needed for a radio programme, you will need to come with good equipment and to use it properly. Unfortunately the fundamental technical changes implied by the introduction of digital audio recording mean that over the next few years choices will be difficult, since expensive equipment could become rapidly redundant. At present you can get the best results on a reel-to-reel machine, recording at no slower than 3¾ inches per second (i.p.s). A high-quality cassette machine can approach this quality for recording (although not for archiving) at a lower cost, but a cheap cassette with a built-in microphone will be hopeless. You will certainly need a separate microphone and some extra money spent on quality in this will be especially worthwhile. Before starting you will probably need to eliminate acoustic problems in the room, carefully setting up the equipment and placing the microphone, which may be pinned on an informant's clothing or even placed like a halter round the neck. Until all this is completed, you need to avoid talking about the subject which you want to record. Although radio producers learn to do all this in a friendly way, commonly with people they have never previously met, there is no doubt that it always adds some tension to the atmosphere. Ordinary historians do not carry the prestige of the media to smooth their requests, nor the funds to buy their equipment, and have little choice but to be content with lesser standards. But that does not mean that it is not worth knowing how to get the best out of the machine that you do have, any more than there is a special virtue in driving badly or typing with two fingers. And there are some elementary rules which will improve the quality of recordings from any tape recorder.

First of all, try to use a quiet room where you will not be disturbed by others talking, and there are no loud background noises, or acoustic problems like those caused by hard surfaces. Traffic outside can be dulled by drawing curtains, but a spitting fire will sound surprisingly loud on the tape, especially if the microphone is not close to the speaker's mouth. In his experience with recording dialect in ordinary homes, Stanley Ellis has found that radio and television, a ticking clock, or a budgerigar, could

spoil a recording completely . . . The acoustics of the room itself should be observed. A tiny room, well stuffed with furniture and with washing airing on a clothes-horse can be an excellent studio. A large quarry-tiled kitchen with plastered walls may give a tremendous reverberation sufficient to spoil the whole recording.[14]

Next, consider where to put the recorder and the microphone. Never place them close together, or you will record the machine's own noise. The recorder is best placed on the floor, out of the informant's view but where you can watch it yourself, and glance from time to time to see if the tape is nearly exhausted without drawing attention to it. The microphone should not be placed on a hard, vibratory surface, nor several feet away from the speaker. Don't record across a hard-topped table. Ideally the microphone should be a foot away from the informant's mouth. With a firm hand, if you choose to sit side by side, you can hold it; or you can place it on a stand, or put it on a cushion or scarf on a side table. All this can be done very quickly. You can stress that it's the informant's voice you need, not the clock or the bird or the radio. And make sure at the same time that the informant is sitting comfortably, and has not given up a favourite chair. Then switch on the recorder and let it run, while chatting. Play this back to test that the recording level is correctly adjusted. If the level is too low, the background noise will swamp the recording; if too high, the sound will become distorted. Then set the recorder running again and, apart from changing tapes, leave it running while the recording session continues. It is a bad practice to keep switching off when the informant wanders off the point, or during your own questions. And never begin with a formal announcement into the microphone, 'This is — interviewing — at — '; it is a formalizing, freezing device. You could leave some spare tape at the beginning to add this afterwards if you wish—but not before, or it may boom out in your playback test.

You are now ready to launch your opening question. What follows will vary greatly depending on the kind of informant, the style of interview you favour, and what you want to know. But again, there are some basic rules. An interview is a social relationship between people with its own conventions, and a violation of these may destroy it. Essentially, the interviewer is expected to show interest in the informant, allowing him or her to speak fully without constant interruption, and at the same time to provide some guidance of what to discuss if needed. Lying behind it is a notion of mutual co-operation, trust, and respect.

An interview is *not* a dialogue, or a conversation. The whole point is to get the informant to speak. You should keep yourself in the background as much as possible, simply making supportive gestures, but not thrusting in your own comments and stories. It is not an occasion which calls for demonstrations of your own knowledge or charm. And do not allow yourself to feel embarrassed by pauses. Maintaining silence can be a valuable way of allowing an informant to think further, and drawing out a further comment. The time for conversation is later on, when the recorder is switched off. Of course you can go too far in this direction, and allow an informant to falter for lack of come-back. To grind to a halt in silence at the end of an exhausted topic is discouraging, and a firm question is needed before this point. But in general you should ask no more questions than are needed, in a clear, simple, unhurried manner. Keep the informant relaxed and confident. Above all, never interrupt a story. Return to the original point at the end of the digression if you wish, with a phrase like 'Earlier you were saying . . .', 'Going back to . . .', or 'Before we move on . . .' But it is axiomatic, if the informant wants to go on to a new line, to be prepared to follow.

Keep showing that you are interested, throughout the interview. Rather than continually saying 'yes'—which will sound silly on the recording—it is quite easy to learn to mime the word, nod, smile, lift your eyebrows, look at the informant encouragingly. You must be precisely clear where the interview has gone, and especially avoid asking for information that has already been given. This demands a quick memory and quite intense concentration. You may find you need to take rough notes as you go along, although it is best to do without this aid if you can. At the same time you should be watching for the consistency of the answers, and for conflicts with other sources of evidence. If you are doubtful about something, try returning to it from another angle, or suggesting, tactfully and gently, that there may be a different view of the matter—'I have heard' or 'I have read that . . .' But it is particularly important not to contradict or argue with an informant. As Beatrice Webb observes pungently:

It is disastrous to 'show off' or to argue: the client must be permitted to pour out his fictitious tales, to develop his preposterous theories, to use the silliest arguments, without demur or expression of dissent or ridicule.[15]

Certainly, the more you can show understanding and sympathy for somebody's standpoint, the more you are likely to learn about it.

Discussion of the past can recall painful memories which still evoke strong feelings, and very occasionally these may distress an informant. If this happens be gently supportive, as you would be to a friend. With some informants it may be wiser to leave the more delicate questions to a later stage in an interview. If it is absolutely essential to get an answer, wait until the end, and perhaps switch off the recorder. But never press too hard when an informant seems defensive or reluctant to answer. It is generally best to try to steer towards a more open conclusion, asking for a summing up of feeling about an experience, or whether anything needs to be added. An interview which ends on a relaxed note is more likely to be remembered as pleasant, and lead on to another.

You need always to try to be sensitively aware of how informants are feeling. If they seem fidgety and are only giving rather terse answers, they may be feeling tired or unwell, or watching the clock for some other engagement: in which case, close the recording session as quickly as possible. While avoiding glancing at your own watch, always fit in with their times, and turn up punctually when you are expected, or they may become tense waiting for you. In normal circumstances, an hour-and-a-half to two hours will in any case be a sensible maximum. An old person, in the interest of the occasion, may not realize the danger of becoming overtired, but will certainly regret it afterwards, and may not want to repeat the experience.

Do not rush away after the recording session. You need to stay, to give a little of yourself, and show warmth and appreciation in return for what has been given to you. Accept a cup of tea if it is offered, and be prepared to chat about the family and photographs. This may be the moment when documents are most likely to be lent to you. It is a good time for fixing another visit. You may find that you can give something in return with some immediate practical help, lifting or fixing something, or some advice about how to set about solving a worrying problem. Indeed, as Ann Oakley has cogently argued, it may sometimes be 'morally indefensible' to hold back from helping in this way, and sharing experience, by talking gently about yourself and your own ideas.[16] Just now and then, this will be the beginning of a friendship which will last. But go forward with tact and caution. Do not get into an argument on subjects likely to be controversial such as teenage behaviour or politics, which is more likely to make for reticence later on.

In some interview situations grander hospitality may be given—an ample lunch with drink—which can perhaps emphasize the normal problem of mutual obligation, bringing a pressure to produce an 'official' version of history. But in most cases you can show sensitivity in making use of the material which has been given, even if it contributes to a conclusion of your own which your informant would not share. Beatrice Webb had no doubts here:

Accept what is offered . . . Indeed, the less formal the conditions of the interview the better. The atmosphere of the dinner-table or the smoking-room is a better 'conductor' than that of the office during business hours . . . A personally conducted visit to this or that works or institution may be a dismal prospect; it may even seem a waste of effort to inspect machinery or plant which cannot be understood, or which has been seen *ad nauseam* before . . . But it is a mistake to decline. In the course of these tiring walks and weary waitings, experiences may be recalled or elicited which would not have cropped up in the formal interview in the office.[17]

Her comment is based on research work of her own in which the normal interview situation was uncommon in two ways: both interviewer and informant were drawn from the top levels of society, and they were roughly the same age. Usually inter-viewers, whether professional historians or the married women typically employed for survey work, are middle-class, and in their thirties or forties. Their informants are normally ordinary work-ing-class or middle-class people, and in oral history work often considerably older. Thus to their normal modesty, or even under-valuation of self, may be added the fragility of old age, and a special vulnerability to discomfort or anxiety. Changing this social balance can have implications for interviewing method which need considering. For example, an interview between the sexes will often help to encourage sympathy and response; but there are some kinds of confidence, for example about sexual behaviour, which are probably much more easily exchanged between married people of the same sex. A very young person, or somebody with a very superior manner, may have more difficulty in gaining trust. Race can provide another kind of barrier. On the other hand, a person from the same working-class background and community as an informant will win an initial rapport, although later on may find difficulty in asking questions because of a common social network, or because the answer (often mistakenly) seems obvious. Similarly, considerable problems of reticence may be encountered if you interview a member of your own family. Differences in social background have to be recognized, and where possible met by variations in interviewing style.

The most recurrent problem is presented by the public personality as informant. Such people are generally tougher and fitter, and perhaps even younger, than the typical informant. They may have such a strong idea of their own story, and what matters in it, that all they can offer is stereotyped recollections. They often also, 'in the course of long careers in public life will have developed a protective shell by which they ward off troublesome questions and while seeming to say something worthwhile in fact give away as little as possible'. This can have become such a habit that 'the subject, even if trying to be frank and open, almost without thinking may reply with the clichéd responses which served so well on other occasions. It is this defensive veil that the interviewer must penetrate.'[18]

Occasionally innocence itself can penetrate the shell. 'Politicians have the right experience to be able to deal very cleverly with a young innocent historian', observes Asa Briggs. But 'a very young man can . . . get a lot from a very old man that members of his own generation don't get'. More usually, there is no alternative but to try to be 'sensitive *and* tough at the same time'.[19] Some of the basic rules still apply: the danger of breaking up the interview through too challenging cross-questioning, and also the positive advantages of, for example, an informal discussion over the dinner-table. Nevertheless, several oral historians, such as James Wilkie in Mexico, Lawrence Goodwin in the southern United States, and Peter Oliver in Canada, have argued for the need to 'cross-examine' in a much more vigorous manner. The oral historian, according to Peter Oliver, while avoiding an openly 'adversary' posture,

should not hesitate to challenge the answers he receives and to probe . . . 'Come on now, Senator, surely there was more to it . . .? Mr So-and-So claims that . . .' Most politicians are pretty worldly and hard-skinned types; few will resent being pushed to re-examine their initial responses if it is done with some tact and skill, and often it is only by doing so that the interviewer will uncover truly significant material.[20]

A comparable instance is provided by the leading radio-astronomers interviewed by David Edge. They combined a very idealized image of science and what was important to its history with the defensiveness needed for success in the competitive grant-aided politics of the scientific world. He developed a triangular method, in which the radio astronomer was interviewed at the same time by Edge, who as a former scientist and perhaps personal friend, and already in possession of inside secrets, was equipped to challenge on technical issues, and by Mike Mulkay, a scientifically naïve sociologist, waiting to pounce on wider inconsistencies and points of interest. David Edge normally led the interview, chasing detailed points, challenging, and arguing; Mike Mulkay came in as an 'outsider', and there was often a notable change in the informant's voice when a question came from him. This argumentative technique clearly depends partly on some sort of common membership of a social group, and partly on knowing exactly how far the challenge may be pressed.

In the extreme reverse situation, the interviewer's chief problems are on quite another, indeed much more basic, level. A European historian collecting oral tradition in Africa is operating in a completely strange culture, and is generally concerned with learning something of its language and basic rules. Among the Kuba, for example, Jan Vansina found that unless all the right people were present, and the right location was chosen, only parts of the traditions of the tribe would be told. 'Among the Akan, sacrifices to the ancestors have to be made before certain traditions are recited, so that the field-worker must be supplied with a sheep or a cask of rum for this purpose.' The Bushongo need to be supplied with home-brewed palm wine, and to recite their tradition at night, in the presence of their ancestral relics. The English historian at home knows not to try to interview a publican on a Bank Holiday or a priest on Good Friday, and can concentrate on less elementary social nuances. Nor does he or she commonly have to rely on interpreters, or to pay for testimonies to be made. Most of the basic rules, such as the avoidance of leading questions, and the need to ensure that the informant is relaxed, apply to the collector in Africa as anywhere; and with ingenuity even some of the special problems can be played against each other:

One should try to see that the informant is . . . prevented from feeling tempted to give false testimony in order to gain favour from the field-workers . . . The informant must not know whether the field-worker is or is not interested in his testimony, for if he does, he will distort it. Hence good informants must not be rewarded more highly than bad ones . . . In addition, during the recording of the testimony, one must adopt a sympathetic attitude towards the informant, without, however, revealing one's real feelings. In Rwanda and Burundi, where I recorded testimonies on magnetic tape, I pretended that I did not understand a word of the language. The clerk who was with me would explain to the informant what he had to do, and then the informant could recite the testimony as he wished. Since he was under the impression that I did not understand what he was saying, he felt that how he said it was unimportant, and he had no particular motive for distorting the tradition.[21]

In one sense this cash-eased exercise in cultural non-communication is as much a parody of how to interview as the worst instances of baiting and insinuating on television back in the ex-imperial capital. One hopes that Africans will before long be creating their own oral history. But these extreme cases do serve to illustrate the need for flexibility in method; and the possibility, too, of securing valuable material in extremely adverse circumstances.

We must return, however, to the ordinary oral historian who was left chatting over a cup of tea. After leaving, three things remain to be done. First, record as quickly as possible any comments of your own on the context of the interview, the character of the informant, additional remarks made off the tape, and what may *not* have been said. Next, label the tape or box. Later on, play back the tape to check what information has been obtained and what you still need. In particular, make sure that you have the basic facts about the informant which any social historian would want to know in order to use it as evidence: the informant's age, sex, home, and occupation, and also his or her parents' occupations. At the same time you can make a list of any names whose spelling needs to be checked with the informant. Finally, if this was your last visit, you can verify these points (providing a stamped addressed envelope again) along with your thank-you letter. This letter can usefully restate the general purpose of the interview, and if appropriate go into questions of confidentiality or copyright. But it is in any case a courtesy which will be valued. And it is on such personal care, just as much as historical expertise, that success in interviewing depends.

# 8

# Storing and Sifting

THE recording has been completed: but how then should the tapes be kept? And how can they be used to make history? We need first to consider the problems of storage and indexing, and then the stages in writing and presenting history with oral evidence.

Because magnetic tape recording is a relatively recent technique, it is far from certain how long it can last and what are the ideal conditions for its storage. The quality of tape has, moreover, been gradually improved, and with this the principal storage considerations have changed. The introduction of digital audio will bring further fundamental changes within the next five years. Good modern tapes no longer have a backing which is likely to disintegrate. The chief problem now is the avoidance of 'print-through', or sound echoes, which can develop during storage. Some experts recommend various means of reducing the risk of print-through, such as running the tape through a recorder once a year so that it is re-spooled, but it is not clear that this is a worthwhile safety measure—indeed, it may on balance create worse risks of other damage. For the moment, there are only two certain rules.

First, the quality of tape to be used for storage should be carefully chosen. If you have used cassette tapes or thin double play reel-to-reel tapes for your original recording, it will be essential to transfer them to standard or long play reel-to-reel tapes (or once it is available, to digital audio tape) for preservation. Otherwise you risk the loss of all your work. It may either become inaudible from print-through, or get stretched or broken when in use on the machine.

Secondly, the place for storing the tape needs to be considered. The tape can be damaged by dust, or by excessive damp or heat. It should never be exposed to temperatures much higher than normal room temperatures by, for example, being stored up against a heating pipe. Modern tapes do not require artificially controlled temperatures or humidity, but the optimum for storage is now considered to be a temperature of 15° to 20°C and a relative humidity of 50 to 60 per cent.[1] Tapes can also be damaged, and even completely wiped of their recordings, by interference from a powerful magnetic dynamo. This risk needs to be taken into account in some buildings, as well as when travelling with them. But in practice, for most oral historians it will suffice to store tapes in a cupboard, wrapped in polythene bags, stood on the shelf in cardboard or plastic boxes on edge, away from the sun or fire or heating pipes, in a room comfortable for working in. And don't smoke or eat near them.

Every tape, as soon as used, needs to be well labelled. It is best to label the box, the spool, and with reel-to-reel the tape itself. The tape can be quite easily labelled on its red and green leads. Without these precautions you may lose tapes by accidentally winding them on to a wrong spool, or replacing them in a wrong box, and then perhaps even making another recording on top of the original. It is, of course, much better if the original tape is kept as a master, and a copy made from it for normal use; and also if a machine is adapted so that it is possible to listen but not to record on it. For a public archive both precautions are essential.

Exactly what you put on the label will depend on how you develop your system of indexing. If you have only a few tapes, it is enough to put the informant's name, 'Tape One, Side One', 'Tape One, Side Two', and so on. Corresponding to this, a box of cards can be kept in alphabetical order, each card with the name of an informant, and a list of the tapes made with them. It is a great time-saver, if you also have transcripts, to note on the cards which pages of transcript cover each side of tape. This box of cards then constitutes an index and catalogue of your collection, and you can easily check whether a tape or transcript ought to be there. The tapes and transcripts can themselves also be kept in alphabetical order to help finding. The disadvantage of this is that each new interview has to be inserted within the existing sequence, rather than added to it. After a while it becomes much easier to store the interviews in order of accession, giving each new informant a number, and adding the number to the index card. If you decide to put only the number on either the tapes or the transcripts, you will also need another index giving the name for each interview number. Similarly, if you decide it is more useful to keep your main index in number order because, for example, this conveniently separates two different parts of your collection, you will still find that you need an alphabetical index which at least gives the number of each informant's interview.

For a small project, one or two boxes of cards along these lines may be all that is necessary. A note of the place and date of recording, made at the time of the interview, can be left as it is with the tape; and the general subject-matter sufficiently remembered to know whether it is worth looking up. But as the collection grows, and especially as more people contribute to making and using it, more information needs to be available in some systematic form.

First, either on the original cards or in a parallel sequence, it is desirable to add to the informant's name, when and where the recording was made, and by whom. It is also useful to note any important variations in the method or quality of recording. If we take a collection of reel-to-reel tapes normally recorded at 3¾, the entry for a somewhat botched recording might look something like that shown in Figure 2 (overleaf).

Secondly, it is worth extracting some of the basic background details about the informant which are essential for evaluating the interview and should thus be found within it. They will of course vary to some extent, depending on the focus of the project. Thus a political collection might include specific entries for elections fought or offices held; and the Imperial War Museum lists details such as 'service', 'arm of service', 'rank', 'decorations and awards', which would be inappropriate in a different context. But most historians need at least to know when an informant was born, his or her parents' occupations, where they lived, whether or not there were brothers and sisters, the informant's own education, occupational career, religious and political affiliation

MRS Sarah JENNINGS          Interview number 36

ADDRESS ...3 Gas Terrace, Woodstock
                    Oxfordshire

Recorded at above/~~elsewhere~~

Interviewer ......Henry Mayhew
Dates of interviews ........31 March 1976 (tape 1)
......and 12 April 1976 (tape 2)

| Tapes | Sides | Transcript pages | Notes |
|---|---|---|---|
| 1 | 1 | 1–16 | |
| | 2 | 16–29 | recorded at 7/8 (ran out of tape) |
| 2 | 1 | 29–45 | |
| | 2 | 45–62 | spoilt by loud hum |

Restrictions of access ...None

Figure 2

if any, whether he or she married, and if so, when, to whom, and whether they had children. All this can again be conveniently summarized on a card, as in Figure 3 (opposite).

All the information can be condensed, and some of it codified, if the suggested form seems too long. At the end of *Speak for England*, Melvyn Bragg has added a very helpful index of 'The People' set out in this form:

160 Joseph William Parkin Lightfoot *b*. Bolton Low Houses 13th December 1908 *br*. Two *s*. Two *pl*. Fletchertown 1938, Kirkland 1942, Wigton 1954 *f.j*. Coalminer *o.j*. Retired, previously coalminer 1922, farm labourer 1924, labourer on pipe-tracks, part-time gardener 1930s, driver Cumberland Motor Services 1942–68, own shop in 1950s *e*. Bolton Low Houses until 14 *r*. Methodist *p*. Labour *m*. Married *ch*. Two

Mrs Sarah JENNINGS                    36

born    1893

at   7 Market Place, Woodstock

father's occupation   bootmaker

mother's occupation   domestic servant before

brothers 3 (one died as infant)    marriage

sisters   2

education   board school till 13

occupations ............................... (dates)....

Assistant in clothes shop              1906-9

apprentice dressmaker                  1909-12

dressmaker in clothes factory          1912-15

part-time outwork from factory at home

politics   inactive                    1915-30

religion   occasional Church of England

lived at ................................. (dates)....

    7 Market Place        1893-1901

    17 Oxford Street      1901-15

    3 Gas Terrace         since 1915


date of marriage   1915

husband/~~wife's~~ occupations

    garage mechanic to c. 1950

        caretaker c. 1950-60

children   1 girl, 2 boys


Figure 3

The abbreviations are self-explanatory, except perhaps for *pl.*, which means 'places lived in'.

A third possibility is to create a series of content cards. For some projects, which are organized to follow a definite interview schedule, this may be superfluous; all the necessary clues will be in the basic background of the informant. But the larger and more diverse a collection, the more a contents card catalogue becomes necessary. One of the most fully developed examples is provided by the BBC Sound Archives. These cards provide a particularly full summary of the contents of each item in the archive, but they begin with a briefer heading. A contents index, depending on the time which is to be spent on it, can aim to be brief or full. But it ought at least to indicate the principal places,

---

CAMPBELL, Beatrice, *Lady Glenavy (Wife of 2nd Baron Glenavy)*          AA

LP28643          D. H. Lawrence And His Circle: the first of two                29.1.64.
                  programmes in which she recalls some impressions of
                  her friendship with
19' 12"          Katherine Mansfield, John Middleton Murry,
                  D. H. Lawrence, and Frieda Lawrence.
                  Producer: Joseph Hone

Copyright: PF                                                      CTIR 38700A
Annots: None
Trans: TP 30.3.64.
Script.

Note: This talk was recorded in Ireland, and is taken from her autobiography
      Today We Will Only Gossip, published by Constable, 9.4.64.

/continued . . .

---

CAMPBELL, Beatrice, *Lady Glenavy (Wife of 2nd Baron Glenavy)*     AA
LP28643                                                        29.1.64.

Recalls first meeting Katherine Mansfield and Middleton Murry, who
were great friends of her future husband, Gordon Campbell:
Katherine's appearance and manner; felt Katherine regarded her as an
interloper into their circle, and tried to shock her with daring
conversation; Katherine's early struggles as a writer; sufferings from
unhappy marriage and love affairs; devotion and care of her friend
Ida Baker; how her hostility to Beatrice overcome by incident during visit
to Paris; the 'psychological dramas' and discussions during evenings at
Parisian cafés.
Gr.90: Through them met Lawrence and his wife, and Koteliansky, known
      as 'Kot'; qualities which made him a friend of Lawrence; Kot's first meet-
      ing with Katherine arising out of quarrel between Lawrence and Frieda,
      and their subsequent friendship; Katherine's association with Murry.

          /continued . . .

---

CAMPBELL, Beatrice, *Lady Glenavy (Wife of 2nd Baron Glenavy)*     AA
LP28643                                                        29.1.64.

Gr.145: Katherine's complex character and varying moods: two occasions
  when she 'put on acts'; a week-end the Campbells spent at the Murry's
  country cottage which was 'not a success'.
Gr.220: Reminisces about time Katherine and Murry spent on visit to
  Campbell's cottage in Ireland; Murry sorry to leave, but Katherine glad
  to return to London.

- 3 -

---

social groups, occupations or industries, political or other
ideologies, personal or family matters, and (more clearly than
these cards) time periods covered.

Finally, especially with a large public collection, it will be
necessary to create a general system of indexing leading to the
other card series. Now that most properly equipped public
archives are computerizing their indexes, up-to-date technical
advice will be needed on the choice of the most suitable program.
It should be user-friendly and flexible: the worst possible mistake
is to get a specialized, tailor-made system individually designed
for the collection, because when its designer goes you will be left
helpless. Key-word-in-context systems are also likely to create
more problems than they solve, because the concepts which will
need indexing can be conveyed by many different words, or indeed
by indirect innuendo, or even avoidance and omission. Since
computers think with rigidly narrow-minded consistency, the key
words will in practice have to be edited into the text before they
can be picked up. This means that with most oral history collections, whether or not using a computer, indexing will be a
process closer to the name and subject indexing of an ordinary
book. Thus all the places, persons, and organizations on the
cards might be included. Important events could be similarly
listed. And with more difficulty, a cross-referenced series of subject headings could be developed. There are at present no clearly

established models to follow, so that it is important to use a system which allows for modification in the light of experience. And above all it should be designed to help, rather than replace, human imagination, understanding, and intuition. In practice this means that the best cataloguing and indexing systems will tell the historian which parts of the collection will repay further investigation, and which will not. Ideally it should be made possible to eliminate, as quickly as possible, all those main sections, or individual items, which are concerned with a different time, place, or general subject-matter from the historian's own interest. Thus before preceding to a contents catalogue as full as that of the BBC Sound Archives, it would be more valuable to break down the general index to this catalogue, so that 'Occupations'—'tin-mining' was further subdivided to lead to 'Cornwall'—'1900–14'; or 'Folk Customs'—'harvest ceremonies' to 'East Anglia'—'1880s'.

Before a recording enters a public archive, another point needs to be clarified, as entries on two of our specimen cards suggest: that of control of the right to access and use. This is not, however, a simple issue, partly because the law of copyright is itself uncertain, varies between different countries, yet has never been properly tested in any of them; but equally because it raises wider ethical questions of responsibility towards informants. The legal position is that there are two copyrights in a recording. The copyright in the recording as a recording is normally the property of the interviewer or of the institution or person who commissioned the interview. The copyright in the information in the recording—the informant's actual words—is the property of the interviewee. But normally some right to use this information is implied by a consent to be interviewed. Thus a person who, knowing that a historian is collecting material for a research study, agreed to be interviewed, would appear to have little ground for complaint if he found himself quoted in print. And in practice he would be very unlikely to attempt to prevent, or to seek compensation for, the publication of any quotation unless he considered it substantially damaging. A bona fide scholar is in fact unlikely to have committed an actionable libel, but it would be foolish anyway to provoke a publicized complaint. It is always important to consider carefully whether the publication of identifiable confidences could not cause local gossip or scandal. Equally, an informant could reasonably complain if information was used in a significantly different context from that suggested; and also, if it proved the making of a best seller, could claim a share of the earnings. If the publication is a single life story, the authorship of the book, and the names on the cover, certainly ought to be decided jointly. For most projects there is much to be said for this balance of rights, and the chief lesson to be learnt is that in explaining a project to an interviewee not only its immediate object, but also the potential value of their information to wider historical research, should be made clear. If the first approach is made in person rather than by post, this can at least be confirmed in a subsequent letter of thanks. An informal understanding of this kind has proved a satisfactory basis for the writing of innumerable sociological studies, as well as most of the oral history publications which we have discussed earlier. Similarly, the fact that in theory some copyright must also exist in most unpublished manuscript material, has rarely produced serious obstacles to the free access of scholars to the holdings of local and national record offices. It may well be that the best policy is normally to leave the issue unresolved. An insistence on a formal transfer of legal rights through explicit, written consent may not only worry an informant, but will actually reduce quite proper protection against exploitation.

There are, nevertheless, contexts in which a formal agreement has become the standard practice. This is the case in broadcasting, where observation of copyright has to be particularly careful because of the frequent involvement of public figures, and also due to the influence of the financial complexities of musical copyright. It is also advised by the Oral History Association of the United States, where standards were originally set for the recording of eminent public figures and a precise agreement was therefore necessary, not only as to copyright, but also as to whether particular pages of the transcript should be closed until a certain date, or accessible only by specific permission. In Britain the Imperial War Museum obtains a precise written agreement from its informants, who are often not merely eminent public figures, but especially security-conscious. The formula advocated in Willa Baum's booklet for American local historians is relatively simple:

I hereby give and grant to the Central City Historical Society as a donation for such scholarly and educational purposes as the Society shall determine the tape recordings and their contents listed below:

(signed) . . . . . (informant).

To this a rider may be added restricting part of the material:

The parties hereto agree that pages 14–16 of the manuscript and the portions of the tape from which these pages were transcribed shall not be published or otherwise made available to anyone other than the parties hereto until 1995.

However, 'except in the few cases where sensitive material is really pertinent, it should be discouraged'.[2]

The Imperial War Museum, which has found that 'it is frequently more difficult to obtain assignments and settle other conditions of deposit and access with executors or heirs than with the informants themselves', seeks a quick exchange of letters 'to tie up all the legal loose ends' along the following lines:

I am now writing to formalize the conditions under which the Museum holds your recordings. The questions which I have already put to you verbally are listed below. I should be grateful if you would let me have your written answers in due course.

1. May the Museum's users be granted access to the recordings and any typescripts of them?

2. May the recordings and typescripts be used in the Museum's internal and external educational programmes?

3. May the Museum provide copies of the recordings and typescripts for its users?

4. Would you be prepared to assign your copyright in the information in the recordings to the Trustees of the Imperial War Museum? This would enable us to deal with such matters as publication and broadcasting, should they arise, without having to make prior reference to you. If you agree to this assignment it does not, of course, preclude any use which you might want to make of the information in the recordings yourself.

Whether or not such a formal agreement is reached, there remains an ethical responsibility towards the informant which is probably more important. First of all, if the recording has been made with an implicit assumption of confidentiality, that must be respected. Any quotation from it which might embarrass the informant must be made either anonymously or with subsequent permission. Similarly, permission should always be sought for its use in a different manner from that originally understood: for example, instead of a history book, for a biographical collection, or a radio broadcast. Moreover, when informants have a right to a royalty fee, as for a broadcast, or a biographical collection, this should be secured for them. They should be warned of the broadcast time well enough in advance to tell friends. And if they are quoted at length in a book, they should receive their own free copy. As far as possible—and admittedly there are some legitimate forms of scholarly publication for which this might be counter-productive—informants' attention should be drawn to the use made of their material. Indeed, an oral historian who does not wish to share with informants the pleasure and pride in a published work ought to consider very seriously why this is so, and whether it is socially justifiable. There may perhaps be a case for publishing the material collected in a more popular form such as a local pamphlet as well as in some academic mode. One accepts that only the outstanding oral historian can reach the range of readership of a Studs Terkel with a single book. But it remains an overriding ethical responsibility of the historian who uses oral evidence to ensure that history is given back to the people whose words helped to shape it.

It should be added that the depositing and preservation of tapes needs to be seen in the same light. They can be of interest and use to far more people than the historian who made the recording. All too many oral history tapes remain with the secretary of a local society, or in an academic's private study, effectively inaccessible to a wider public. This may be reasonable while they are being actively used for personal research, but commonly continues beyond this, partly because national and county record offices have only slowly organized facilities for storing and listening to tapes. But the offer of the original tapes, or copies, to a local record office or a public or university library, besides being desirable in itself, may stimulate the provision for those needs, and prove the seed for a significant collection—an asset which will find many different uses within the community.

For the same reason there is a strong argument, whatever the immediate use envisaged for them, for the full transcription of tapes as the first stage in the writing and presentation of history. Transcribing is undoubtedly very time-consuming, as well as being a highly skilled task. It takes at least six hours, and for a recording with difficult speech or dialect up to twice as long, for each hour of recorded tape. Yet unless the tape is fully transcribed, anybody but the person who made the recording—and so has quite a clear idea of what it contains—will be severely hampered in using it. A contents card is at best only a rough guide for the visiting researcher: listening to more than a few tapes takes several hours, where skimming through transcripts might take minutes. But the person who makes the tape is also best able to ensure that transcription is accurate. Because this task is so lengthy, and, apart from other claims on time, recording always seems more urgent, transcribing nearly always falls behind. In a research project supported by a grant, this can be avoided only by making a full estimate of the transcribing time and equipment needed at the start. Allowance ought to be made for headphones, so that the transcriber is not distracted by background noise, and for a tape recorder with a reverse foot-pedal for play-back: both are essential for transcribing at a reasonable speed. It will also reduce the time needed by roughly a quarter if the transcripts are typed on an electric typewriter with a memory or a personal computer or word processor, which now costs no more. With a personal computer you can also keep your transcripts in disk form so that they can be edited later: but allow for the cost of the disks too.

It is equally important to recognize that transcribing work can only be carried out by a person with particular skills, working on a regular basis. Part-time agency audio-typing will either be incomprehensible or prohibitive. A transcriber needs to be interested in the tapes, intelligent in making sense of them, especially in the key art of turning verbal pauses into written punctuation, and a good speller with an unusually quick ear. It is also isolated work. These are not necessarily the qualities which make a successful secretary. The only way to know whether somebody can transcribe well is to give them a tape and let them try.

Most oral history projects will not have the resources to pay for a transcriber, and will need to carry out the work themselves. For a very small group, or for a researcher's own tapes, the process can, however, be quite markedly shortened, even if at the expense of long-term satisfaction. The best 'shortened transcript' lies between the full contents card and the complete transcription. For the most part, the content is summarized in detail, but actual quotations are only used when the words are so well or vividly put that they are worth considering for extracts or quotations in the finished presentation. A finding device can be added in the margin, either by using the numbers in the counter-setting on the machine (although these unfortunately vary even between machines of the same make), or by listening through the tape after transcription and noting the time intervals every five or ten seconds (standard, but less quick to use).

Ultimately, however, there can be no substitute for a full transcript. Even the best shortened version is like an intelligent historian's notes from an archive rather than the original documents. Nor can the historian today know what questions will be asked by historians in the future, so that any selection will result in the loss of details which might later prove significant. The full transcript should therefore include everything, with the possible exception of diversions for checking that the recorder is on, having a cup of tea, or present-day chatting about the weather, illness, and so on. All questions should go in. Fumbling for a word may be left out, but other hesitations, and stop-gaps like 'you know' or 'see', should be included at this stage. The grammar and word order must be left as spoken. If a word or phrase cannot be caught, there should be a space in the transcript to indicate this. These are all quite straightforward guidelines. But the real art of the transcriber is in using punctuation and occasional phonetic spelling to convey the character of speech.

The transcript is in this sense a literary form and the problems which it raises are inseparable from those of subsequent quotation.

The spoken word can very easily be mutilated in being taken down on paper and then transferred to the printed page. There is already an inevitable loss of gesture, tone, and timing; and some deliberate changes will be needed to cut out pauses and distracting hesitations or false starts in the interests of readability. Much more serious is the distortion when the spoken word is drilled into the orders of written prose, through imposing standard grammatical forms and a logical sequence of punctuation. The rhythms and tones of speech are quite distinct from those of prose. Equally important, lively speech will meander, dive into irrelevancies, and return to the point after unfinished sentences. Effective prose is by contrast systematic, relevant, spare. It is therefore very tempting for the writer, wishing to make a point effectively, to strip a spoken quotation, re-order it, and then, in order to make it continuous, slip in some connecting words which were never in the original. The point can be reached when the character of the original speech becomes unrecognizable. This is an extreme, but any writer, unless continually aware of this danger, may at times reach such a level of decadence in transcription.

The difficulties may be illustrated by taking as an example one of the first passages in Ronald Blythe's *Akenfield*, an old farm worker's account of a domestic economy in the years before 1914. The picture he gives is very bare, highly effective—but so terse in detail that one wonders how far the original interview has been tidied up:

There were seven children at home and father's wages had been reduced to 10s. a week. Our cottage was nearly empty—except for people. There was a scrubbed brick floor and just one rug made of scraps of old clothes pegged into a sack. The cottage had a living-room, a larder, and two bedrooms. Six of us boys and girls slept in one bedroom and our parents and the baby slept in the other. There was no newspaper and nothing to read except the Bible. All the village houses were like this. Our food was apples, potatoes, swedes and bread, and we drank our tea without milk or sugar. Skim milk could be bought from the farm but it was thought a luxury. Nobody could get enough to eat no matter how they tried. Two of my brothers were out to work. One was eight years old and he got 3s. a week, the other got about 7s.[3]

There is in these lines an unremitting logical drive. Every word stands with evident purpose in its proper place. Every phrase is correctly punctuated. There are no ragged ends, no diversions to convey the speaker's own sense of a childhood home, or the bitterness or humour felt in poverty. Some phrases read like the author's own comments: 'skim milk . . . was thought a luxury.' There are no dialect words, no grammatical irregularities, no sparks of personal idiosyncrasy. The passage may convince but, unlike many others in the same book, it does not come alive. One wishes to know, but is provided with no indication of, where the interview has been cut, and what has been put in to sew it up again.

We can turn for a contrast to George Ewart Evans's *Where Beards Wag All*, also about Suffolk villagers, some of them from the same community. This is a book with more direct argument than *Akenfield*, but supported by substantial quotations in which we seem to hear the people themselves talking, even thinking aloud, in their own, very different style, as this old man:

It's like this: those young 'uns years ago, *I said*, well—it's like digging a hole, *I said*, and putting in clay and then putting in a tater on top o' thet. Well, you won't expect much will you? But now with the young 'uns today, it's like digging a hole and putting some manure in afore you plant: you're bound to get some growth ain't you? It will grow won't it? The plant will grow right well. What I say is the young 'uns today have breakfast afore they set off—a lot of 'em didn't used to have that years ago, and they hev a hot dinner at school and when they come home most of 'em have a fair tea, don't they? *I said*. These young 'uns kinda got the frame. Well, that's it! If you live tidily that'll make the marrow and the marrow makes the boon (bone) and the boon makes the frame.[4]

We have to pause here to listen, accept the difficult rhythm and syntax of his speech, ruminating, working round to the parable image which he has held all the time in store. This quotation certainly requires more adaptation by the reader. But that may be needed, and if so will become generally learnt, as the qualities of speech become more understood.

George Ewart Evans is using artistry in his quotation as much as Ronald Blythe. Probably some hesitations, pauses, or repetitions have been eliminated from the recorded speech, and he has put in punctuation. But he has done this in a way which preserves the texture of the speech. Italics are used to indicate unexpected emphasis, and punctuation to bring the phrases together rather than to separate them. The syntax is accepted; the breaks in the passage left. And occasionally a word is spelt phonetically to suggest the sound of the dialect. Too much phonetic spelling quickly reduces a quotation (from whatever social class) to absurdity, but the odd word to convey a personal idiosyncrasy, or a key tone in a local accent like the Suffolk 'hev' and 'thet' used here, help to make a passage readable as speech without losing any of the force of its meaning.

In transferring speech into print the historian thus needs to develop a new kind of literary skill, which allows his writing to remain as faithful as possible to both the character and meaning of the original. This is not an art normally needed in documentary work. But the analogy with documentary quotation in other ways sets a useful standard. It is unfortunately not the usual practice in sociological studies quoting interviews to indicate cuts and other alterations. Historians can, however, insist on the care normal in their own discipline, showing excisions by a dotted line, interpolations by brackets, and so on. A re-ordering cannot be acceptable if it results in a new meaning, unintended by the speaker. And the creation of semi-fictional informants, by exchanging quotations between them, or dividing two from one, or creating one out of two, must always be by the standards of scholarship indefensible. An oral documentary which does this may gain in effect, but it becomes imaginative literature: a different kind of historical evidence.

Oral historians in the United States have introduced an additional standard in their practice. After transcription, typescripts are sent to the informant for correction.[5] This clearly has advantages in picking up simple errors and misspellings of names. It can also result in stimulating new information, and political historians who use the interview method often send transcripts for this purpose. But it has drawbacks too. Many informants find it impossible to resist rewriting the original conversational speech into a conventional prose form. They also may delete sentences and rephrase others to change the impression given from some particular memory. Since the original tapes are rarely consulted in American archives, and the transcript rather than the tape is regarded as the authoritative oral testimony, the process of correction weakens the authenticity of oral evidence in use. In addition, while some informants, like retired public figures, may have the time and confidence for correcting a lengthy transcript, there are probably many more for whom it would simply be a worrying imposition. For these it is much better to write asking only for a few clarifications of confusions, uncertain names, or vital details missing—which will usually be gladly supplied.

With transcription started, the sorting of the material for use can also be begun. It is best to make at least three copies of the transcript—and a fourth, if one is to go to an informant. The top copy can then be filed as a complete interview, a series parallel to the tapes themselves. The other copies can be re-sorted, and divided or cut up into different subject files (the third copy being used for cases where subjects overlap), depending on what use is in mind. Whole interviews could be put together by place, by social group, or by occupation. Alternatively, the passages within each interview about school, or church, or family could be cut out (marking the page in the transcript from which they are taken), and placed in a series of boxes. These boxes may well follow the sequence of the original schedule of questions. Then if a question has been asked, for example, about church attendance or how people met their husbands or wives, all the relevant material can quickly be found together in the same box. But the precise choice of method in re-sorting must depend upon the form of analysis and presentation intended. It is to this essential, final question that we must now turn.

# 9

# Interpretation: The Making of History

THE evidence has now been collected, sorted, and worked into an accessible form: the sources are at our disposal. But how do we put them together? How do we make history from them? We shall have to consider, first, what choices may be made in the manner and shape of presentation. Next, how do we evaluate and test our evidence? Thirdly comes the heart of the matter, interpretation: how do we relate the evidence we have found to wider patterns and theories of history? How do we construct meaning in history? And finally, we shall take a concluding look ahead, at the impact which we might hope for from oral evidence on the making of history in the future.

The presentation of history with oral evidence opens new possibilities. As a whole, as we shall see, the essential skills in judging evidence, in choosing the telling extract, or in shaping an argument, are much the same as when writing history from paper documents. So are many of the choices: between, for example, audiences of other historians, of schoolchildren, local newspaper readers, or an old people's club. Oral history does, however, highlight the need for some of these choices, simply because it can be effective in so many different contexts.

The first choice is the medium, for its techniques and conventions will shape and limit the message which can be conveyed. In the future it may become easier to combine both sound and print, issuing a tape of excerpts, for example, to accompany a book. It is already common enough for booklets and printed programmes to be sent out as an aid to broadcasting. Nevertheless, for the moment oral history is normally presented in one of a number of separate forms.

The first is the radio broadcast; sound only. There is a whole range of possibilities here, from the raw material itself in an autobiographical interview to the illustrated academic talk.

Broadcasting has also developed a very special art of conveying scenes and messages in sound. Original tapes can be not merely clarified by cutting out hesitations and pauses, but heightened by rearrangements of words; and background noises can be inserted. Some of this amounts to a tampering with evidence which a historian ought not to accept, but the fine editing of a sound tape which becomes possible with the technical resources of broadcasting can certainly make quotation briefer and more effective. Sound can also make some clues superfluous, so that a series of extracts in different regional or class accents can be directly juxtaposed. Indeed a whole programme can be designed as a collage of sound, with very little or no connecting narrative at all, and 'footnotes' perhaps supplied by the programme notes. In this way a historical picture of a community can be built up, such as a fishing town, interweaving the sounds of the herring gulls and the auctioneers at the quay, with old people's accounts of how the men caught fish, how the women gutted it and mended nets, stories, singing in the pubs, hymns, and preaching in church.

When pictures are also added in broadcasting, with television, there is a radical shift in what can be conveyed. The visual effects tend to dominate. Fine cutting is not possible in an interview unless a separate visual sequence is introduced, because otherwise there is likely to be a jump in the interviewee's physical position at each cut. But a separate visual sequence is distracting, conveying its own meanings. The same problems apply with a collage. Since the verbal messages conveyed can be less easily clarified, and the pictorial meanings tend to be symbolic rather than precise, television presents argument in a more diffuse form than radio. But seeing the informants themselves, and old photographs of their families, homes, and workplaces, does bring another dimension of historical immediacy.

In a more elementary way sight and sound can be combined in tape and slide shows for many kinds of historical presentation, from reminiscence group therapy to a formal illustrated lecture. Tapes of excerpts are already available from many museums and libraries as part of their schools service. You can also make your own. The simplest way of using them is in a talk whose main purpose is to arouse interest: a short introductory explanation, followed by extracts. Since the accent may be a little hard to catch at first (and the tape recorder very likely does not amplify well enough) it is best to choose a few, clear, quite long extracts—four or five minutes each. It will help if you supply the audience with duplicated copies of the transcripts. For a more elaborate lecture, where extracts are used to illustrate an argument which may be quite complex, this solution is less easy. In this case you must have good recordings in the first place, and they will need to be copied from the originals on to a single tape of extracts.[1] You must also make sure that a reliable amplifying system is available in the lecture room. You can then lecture standing next to the play-back machine, flicking the quotations on and off with the pause lever. Without such preparation, as most oral historians by now know to their cost, an audience is likely to be puzzled by incomprehensible voices, distracted by intervals while the right place is found on the tape, and irritated by a gross over-running of time.

A second choice which arises naturally from the origin of oral evidence in the co-operation of an interview—and very often the carrying out of field-work by a group—is the possibility of a jointly edited publication. Indeed, in presentation through radio or television, teamwork is of course essential. Here the roles are clearly defined: technicians, producer, historian, interviewee. But with printed publications a more flexible approach is possible. For a school project, or a community oral history, the collective work in putting together oral material may be as valuable an experience as the recording itself. In a community project a group of old people may record each other's reminiscences, discuss them together, decide what to choose for publication, correct and elaborate the scripts, and so on. In a school project the co-operation will be more likely over production: choice of the best extracts, design, and printing.

Equally basic is the need in almost any form of presentation to decide between approaching history through biography or through a wider social analysis. Oral evidence, because it takes the form of life stories, brings to the surface a dilemma which underlies any historical interpretation. The individual life is the actual vehicle of historical experience. Moreover, the evidence in each life story can only be fully understood as part of the whole life. But to make generalization possible, we must wrench the evidence on each issue from a whole series of interviews, reassembling it to view it from a new angle, as if horizontally rather than vertically; and in doing so, place a new meaning on it. We are thus faced with an essential but painful choice.

There are broadly three ways in which oral history can be put together. The first is the single life story narrative. For an informant with a rich memory it may well seem that no other choice can do the material full justice. Nor need a single life narrative present just one individual biography. In outstanding cases it can be used to convey the history of a whole class or community, or become a thread around which to reconstruct a highly complex series of events. Thus the autobiography of Nate Shaw in *All God's Dangers* is powerful just because it stands for the wider experience of the Black people of the southern United States. A story of such power asks for no more than a brief explanation of its context; others, especially if intended to be read as in some sense typical, will require a much fuller introductory discussion and interpretation if they are to reach beyond the anecdotal.

The second form is a collection of stories. Since none of these need be separately as rich or complete as a single narrative, this is a better way of presenting more typical life-history material. It also allows the stories to be used much more easily in constructing a broader historical interpretation, by grouping them—as a whole or fragmented—around common themes. Thus Oscar Lewis explores the family life of the Mexican city poor in *The Children of Sanchez*, by taking for one family the different accounts of parents and children and bringing them together into a single multi-dimensional picture. On a larger scale, a group of lives may be used to portray a whole community: a village, as in *Akenfield*, or a town, as in *Speak for England*. Or the collection may focus upon a single social group or theme, like *Fenwomen*, or *Working*, or *Blood of Spain*. It may be organized as a collection of whole lives, or stories about incidents, or as a thematic montage of extracts: *Blood of Spain* interweaves all three. And here again the character of the introduction will also shape the impact of the stories.

The third form is that of cross-analysis: the oral evidence is treated as a quarry from which to construct an argument. It is of course possible within one book to combine analysis with the presentation of fuller life stories. In my own *Edwardians*, a series of family portraits, chosen to represent the varied social classes and regions of Britain, is interwoven between the more directly analytical chapters. But wherever the prime aim becomes analysis, the overall shape can no longer be governed by the life story form of the evidence, but must emerge from the inner logic of the argument. This will normally require much briefer quotations, with evidence from one interview compared with that from another, and combined with evidence from other types of source material. Argument and cross-analysis are clearly essential for any systematic development of the interpretation of history. On the other hand, the loss in this form of presentation is equally clear. Because of this, these basic forms are not so much exclusive alternatives as complementary, and in many cases the same project needs to be brought out in more than one of them.

The chosen form will itself partly determine how sharp the distinction will be between oral and other documentary source material in presentation. It is least obvious in written forms. The problems of transcription will need to be considered, and a system for citing interviews chosen. After writing, the manuscript ought to be checked against the original tapes, a task which will only be difficult if they are not transcribed. And the material has to be interpreted with a full awareness of the context in which it was collected, the forms of bias to which it is liable, and the methods of evaluation which are thus needed: matters which are our next concern. Above all, and always most challenging, to succeed fully as history an integration must be achieved of generality and detail, of theory and fact.

Writing a book which uses oral evidence, either alone or with other sources, thus does not in principle demand many particular skills beyond those needed for any historical writing. The oral evidence can be evaluated, counted, compared, and cited along with the other material. It is no more difficult, no easier. But in some ways it is a different kind of experience. As you write, you are aware of the people with whom you talked; you hesitate to give meanings to their words which they would wish to reject. Humanly and socially, this is a proper caution; and indeed anthropologists have shown it equally essential to scientific understanding. In writing, too, you strongly wish to share with others the insights and vividness of the life stories which have held your own imagination. Moreover, this is material which you have not just discovered, but in one sense helped to create: and is thus quite different from another document. This is why an oral historian will always feel a specially strong tension between biography and cross-analysis. But this is a tension which rests on the strength of oral history. The elegance of historical generalization, of sociological theory, flies high above the ordinary life experience in which oral history is rooted. The tension which the oral historian feels is that of the mainspring: between history and real life.

The next stage is the evaluation of the material which has been collected. We have earlier considered at some length, in the chapter on Evidence, the forms of bias to which oral sources are subject, and how far these are shared with documentary evidence. But how in practice does the historian evaluate oral source material?

There are three basic steps which must be taken. First, each interview needs to be assessed for internal consistency. It must be read as a whole. If an informant has a tendency to mythologize or to produce stereotyped generalizations, this will recur throughout an interview. The stories in it may then still be taken as symbolic evidence of attitudes, but not as reliable in factual detail as they might be with another informant. Similarly, suppression of information can be revealed by a repeated avoidance of discussion of a particular area—or through unresolved contradictions of detail (such as date of marriage, and the birthdate and later age of a first child, which was conceived before marriage). Any extensive suppression or invention will produce extremely obvious inconsistencies, contradictions, and anachronisms, especially if the interview takes more than one session. In such a case, it is best to discard the entire interview. On the other hand, some inconsistencies are quite normal. It is very common to find a conflict between the general values which are believed true of the past and the more precise record of day-to-day life; but this contradiction can be in itself highly revealing, for it may represent one of the dynamics of social change—and a perception which is, in fact, rarely possible through any other source than oral evidence. On a more humdrum level, memory is in general less precisely reliable on a matter of chronology, or a brief once-for-all incident, than on the detail of a recurrent process of work or social or domestic life. By contrast, a small minority of informants can be found whose richness and consistency of memory is absolutely exceptional. Because it is so extensive, the accuracy of such a memory is easier to confirm from other sources: a list of land occupiers, for example, from the rate books; or the year of a local suicide might be traced back to a newspaper report. But even with such cases, as with others, by first looking at the interview as a whole, you can arrive at a good measure of the general reliability of the informant as a witness.

On many points a cross-check can be made with other sources. This can of course be a cumulative process as material is gathered in: a series of interviews from the same locality will provide numerous factual cross-checks between each other. Details can similarly be compared with manuscript and printed sources. '*Any* evidence,' as Jan Vansina puts it, 'written or oral, which goes back to *one* source should be regarded as on probation; corroboration for it must be sought.'[2] This dictum may, however, be of more general relevance to oral tradition handed down over several generations than to direct life story evidence. Where there are discrepancies between written and oral evidence, it does not follow that one account is necessarily more reliable than another. The interview may reveal the truth behind the official record. Or the divergence may represent two perfectly valid accounts from different standpoints, which together provide vital clues towards the true interpretation. Very often, indeed, while oral evidence which can be directly confirmed proves to be of merely illustrative value, it is fresh but unconfirmed evidence which points the way towards a new interpretation. Indeed, much oral evidence, springing from direct personal experience—like an account of domestic life in a particular family—is valuable precisely because it could come from no other source. It is inherently unique. Of course its authenticity can be weighed. It cannot be confirmed, but it can be judged.

The third method by which such a judgement can be reached is by placing the evidence in a wider context. An experienced historian will already have learnt enough from contemporary sources about the time, place, and social class from which an interview comes to know, even if a specific detail is unconfirmable, whether as a whole it rings true. General absence of reliable detail, anachronistic attitudes, and incongruous linguistic phrasing will all be obvious enough.

It is possible with more special techniques to push further than this. For example, an expert in dialect may be able to identify exactly the extent to which an informant has kept or modified the local vocabulary of his or her birthplace. Or a folklorist may be able to pick out stories which are versions of known tales, distinguishing the elements in them which are unaltered and those which are new. The whole interview may, in fact, be read, or listened to, in this spirit, as a piece of oral literature. Although still insufficiently developed in relation to the typical oral history interview, a form of literary analysis is the next step which can be taken in the interpretation of the material. This could follow one of several different approaches.

First the historian may seek to understand an interview in the sensitive, humanistic manner of the traditional literary critic who interprets the meanings intended by the author, often in a confused and contradictory text, from all the clues in it which seem helpful. Thus Ron Grele contrasts two interviews, each with a working-class Jewish New Yorker from the tailoring trade. Despite their similar backgrounds, they present history in fundamentally different ways. For Mel Dubin, an immigrant's son born in the city, skilled worker and union organizer, history is an uphill struggle for progress, chronological, and despite its setbacks, logical. In each dimension of his account—his personal story, the neighbourhood, the union, the garment industry—he constructs the same pattern of rise and decline, and gives the same explanation, the disappearance of the skilled immigrant Jewish and Italian tailors of earlier decades: just the skills on which Mel's own life was built. Mel's history, constructed from both direct experience and knowledge of the past, and also with the help of significant omissions and exaggerations, is a historical myth of progress, 'which functions in very particular ways to give a dynamic to the tale, and leads inevitably to certain very real conclusions about the nature of the world of the garment industry today'. Bella Pincus, on the other hand, also a militant, was herself an immigrant, coming to the city as a teenager from a village in Russian Poland; she worked before marriage as a semi-skilled machine operator, and returned again to work as a widow. Bella does not present history as the logic of change, but as a series of dramatic episodes, all exhibiting the same moral lesson of struggle: 'It's always the same. Ever since the world is it's rich and poor, struggling and well off. That's how it is.' It is indeed closer to her own history. And she tells it with the constant poetic use of paired images. For example, she describes her first impressions of New York in terms of the open-top buses, the flat roofs of the houses, and the washing in the street, in contrast with the closed buses, pitched closed roofs, and hidden washing of her Russian childhood: symbols which also give the sense of openness she felt in her own life when she was a young girl in New York, compared to her life in Russia, and to her life now. Thus in both these life stories it is not only through the facts and opinions given, but perhaps still more through the imaginative and narrative skills with which they are put together, that we can perceive the speaker's deeper historical consciousness. This is all the more striking since they had to fight to be heard in the interview, returning 'again and again to the main thrust of his or her story, despite the sometimes strained efforts of the interviewers to control the situation and to divert them to other questions'. The need to 'Listen to their Voices', both in the interview and afterwards, is here put conclusively.[3]

Luisa Passerini has found a parallel contrast in a group of interviews with Turinese workers between a minority—mostly men—who portrayed their lives in terms of choice, decision, acquiring skills, searching, and sacrifice, and the majority who spoke of themselves as fated, 'born socialists', born rebels, born to poverty, and so on. She sees these messages, however, as often not consciously intended, but reflections of the ideas in an earlier, archaic popular culture surviving in spoken language: as for instance with the woman who explains her childhood pranks, her marriage without the permission of her parents, and her insistence on being a working wife, by saying 'I had the devil in me'.[4]

Such half-conscious meanings can also be discerned in the formal qualities of language itself. Written language is grammatically elaborate, linear, spare, objective, and analytical in manner, precise yet abundantly rich in vocabulary. Speech on the other hand is usually grammatically primitive, full of redundancies and back-loops, empathetic and subjective, tentative, repeatedly returning to the same words and catch-phrases. But these contrasts are not absolute within either speech or writing: there are marked differences between individuals in vocabulary and grammar, tone and accent, which reflect regional origin and education, social class and gender. Thus in European literary writing up to the nineteenth century, because they were usually better educated, men adopted a more rhetorical, academic, formal style than women. But when ordinary people tell life stories, the men are more likely to use the direct, active, subjective mode, the 'I', and women the indirect, reflective 'we' or 'one'.[5] And choices of particular key words and catch phrases, for instance when conveying moral attitudes, will again vary, both between speakers and with the same speaker in different contexts, and can be equally telling of assumptions, often unspoken, and sometimes deeply buried.

These hidden meanings can be read without accepting the view of some recent linguistic and psychological theorists that grammar itself moulds the infant consciousness. Others have similarly seen narrative as the primary form through which human beings make sense of their experience. Certainly unsolicited, often ironical and humorous stories are a recurring device for conveying symbolic messages, both in interviews and in real life. Carolyn Steedman's father could not tell her that he had not married her mother, or discuss the arrangement by which he paid her mother an allowance and kept a workroom in the house while living with another woman; yet in his hoodwinking of the new lodger on the stairs with his greeting, 'Hello, I'm the other lodger', an oft-remembered incident which became 'our only family joke', he also told everything—in a nutshell.[6] Unfortunately there is little to guide the historian in the analysis of stories and jokes. Thus the structural analyses of Black English vernacular by William Labov certainly show the story-tellers' technical skill, but offer few clues for the symbolic interpretation of their messages; and while Luisa Passerini has again drawn suggestively from the special knowledge of popular stories and song in folklore, discussion in that field has too often been in terms of a traditional past which most historians regard as a myth in itself. The chance to develop a new method appropriate to oral history remains open.

Yet another possibility is to examine the interview as a literary 'genre', imposing its own conventions and constraints on the speakers. Thus Robert Fothergill has traced the evolution of English diaries from the journal of daily events and the Puritan diary of conscience, to the private reflective diary which only became the accepted genre from the late eighteenth century. David Vincent has shown how stylistic difficulties in part explain why early nineteenth-century working-class autobiographers wrote freely about their public lives but rarely about intimate feeling. But comparisons between different kinds of personal document, including interviews, have yet to be carried out in English. Luisa Passerini found some active Catholics and also socialist militants adopted a form of life story similar to that used for saints, sometimes referring to this 'self-hagiography' as 'my confession'. And Stefan Bohman has made an especially suggestive comparison of diaries, written memoirs, and interviews from the same Swedish working men. He found the diaries, small printed pocket-books crammed with tiny writing, still traditional journals of events, mainly about the weather and work: none took up the form of the private reflective diary. Memoirs and interviews were more alike, using the same stories and even phrases, but with important differences too. The written memoirs focused on early life, and they used a more public, abstract language. Thus one man writes:

My father died in Stockholm on 2nd August 1933. He died in extreme poverty after a long illness, patiently borne. What can I have done to deserve such suffering he said—and poor Mother.

He even uses the stock phrase of public memorial announcements, 'after a long illness patiently borne'. His account in the interview is much more personal and detailed—and as a result, significantly different in what it conveys:

Yes, he died at home. I came home one day the last year, I came home when I was out of work. He was lying there on an iron bedstead. We were incredibly poor. It was in the afternoon, three or four o'clock. I saw there was blood and a bloodstained handkerchief on a chair by the bed . . . He'd taken a razorblade and cut both his wrists, hacked at them. But he hardly bled at all, he was so thin. He thought he was a burden on the family.

   'What have I done to deserve to suffer like this?' he said.

Another man writes of his own later years in his memoir:

As a result of the conditions now prevailing, the summer home or my 'life's work', if I may thus express myself, has if anything become a burden in the financial sense, now that I have retired. Unless I sell it, which I do not wish to do. I am not altogether satisfied with the taxable values stated above. I regard this as a construction making it possible to collect money from a hard-working and perhaps somewhat naïve citizen.

What he really wants to say is trapped in the conventions of the style of writing which he believes appropriate for a public memoir. The interview releases the same message in quite a different manner:

Djurö is my life's work. I worked and slogged and hammered away like hell, and I scraped and saved to do it. But the taxable value, you see, it's hard to keep it on, I could sell it if I wanted to . . . It's rather a smack in the teeth. Somebody in the same position could have blown the money and lived it up. Then he can go on the welfare. I've never got a penny-piece that way, and I don't want to either.[7]

Philippe Lejeune has also compared various forms of autobiography in France, evaluating a whole series of different genres including the autobiography in the third person, the radio interview, the 'document vécu', and the oral history interview. He is particularly illuminating in his discussion of the modern 'document vécu': the candid autobiography 'from the horse's mouth', which reveals the hidden story of a prison or hospital, a murder or sexual scandal, of war or resistance, or simply the unknown lives of ordinary people like peasants or fishermen, which French publishers have brought out in series with titles such as 'Témoignages', 'Elles-mêmes', or 'En direct'. He shows these are shaped partly by opposition to other forms: the nurse's own experience, for example, is an answer to romantic hospital novels with doctors as the heroes—'the men in white'—and also to the official literature of her own profession. More generally, they are assumed to contrast with the self-conscious literary autobiography, and presented as direct, readable, even artless: but in practice they repeatedly use the same devices, such as the present tense, the diary form, and direct dialogue, and are organized dramatically as a clear story told through a series of scenes. Tantalizingly, Lejeune fails to follow this with a comparison of the forms and devices found in oral history interviews, but reverts to the question of dual authorship and its precedents in the ghost writers of earlier autobiographies.

This itself leads to a last form of literary analysis. Elliot Mishler has cogently argued that the interview should be interpreted as a joint product between *two* people, 'a form of discourse . . . shaped and organized by asking and answering questions'. His own experience comes from medical interviews, where the asymmetry of power between questioner and respondent is especially marked, because it is only the former who has favours to offer: the right information makes a cure more likely. Mishler shows how quickly the patient keys in to the doctor's responses—either meaningful silences or requests for more detail—and cuts out circumstantial comment, often ending up with simple 'yes' and 'no' answers. He alerts us to the need to watch the questions as well as the answers in interpreting an interview. In 'the mainstream tradition' of social science surveys this mutual exchange of meanings is suppressed, both at the interview stage and also later in the process of coding, but with recorded evidence there is the chance to examine the whole dialogue.[8] Unfortunately there are few practical examples to follow. Among the experts who have studied texts from this perspective, symbolic interactionists and hermeneuticists seem absorbed in proving that there is indeed a mutual dialogue, but fail to go on to interpret it. Literary structuralists have concentrated on the formal stylistic filters in communications between people, to the extent that some seem trapped in an impossibilism. They see both speakers drawing on a full repertoire of inflection, tone and gesture as well as words, yet unable either to take all this in or to convey a clear message in the first place; rather, the speakers are 'expressing a plenitude of meanings, some intended, others of which [they are] unaware'.[9] From such positions there is no logical advance: just old-style intuitive guesswork. Worse still, too often these theories are phrased in deliberate obscurity, self-referring in their complexity. Trapped in webs of scholastic 'discourse', it is easy to forget the important messages that do get across—even by telephone, or telegram, or between people who speak different languages: to forget that the informant had something to say; in short, to stop listening. In recognizing the interview as 'a form of discourse', we must not forget that it is also a testimony.

Interviews, like all testimonies, contain statements which can be weighed. They weave together symbols and myths with information, and they can give us information as valid as that obtainable from any other human source. They can be read as literature; but they can also be counted. To start with, a group of interviews can be tested to see how the basic information they contain measures up against that known from other sources. Thus in his study of 'The Family and Community Life of East Anglian Fishermen', Trevor Lummis tabulated some of the information collected from sixty interviews.[10] Informants were asked the age at which they left school. Their answers fit neatly with known national trends, both with time and across social class:

| % left school | Born | | | Son of | | |
|---|---|---|---|---|---|---|
| | before 1889 | 1890–9 | 1900–9 | owner | deep-sea skipper | deep-sea crewman |
| at 11 or 12 | 36 | 15 | 7 | 0 | 16 | 33 |
| at 13 | 53 | 33 | 36 | 22 | 69 | 33 |
| at 14 or 15 | 11 | 52 | 57 | 78 | 15 | 33 |

Information had also been collected on the number of informant's brothers and sisters, and whether any died in childhood.
Fishermen are known to have been unusually slow in reducing
family size. When tabulated, the figures again prove compatible
with national trends towards lower infant mortality and fewer
children—as again they are with known differences between
social classes:

| | Born | | | Father | | |
|---|---|---|---|---|---|---|
| | before 1889 | 1890–9 | 1900–9 | owner | deep-sea skipper | deep-sea crewman |
| Number of brothers and sisters | 9.9 | 7.0 | 7.9 | 9.1 | 8.5 | 9.5 |
| % who died as children | 15 | 14 | 7 | 11 | 15 | 25 |

The historian with such test results at his elbow can move forward
with some confidence into less charted terrain.

At this stage, some will be looking for patterns, clues towards
interpretation, in the facts before them. Others will have started
from a more definite theoretical standpoint, and probably some
more detailed lesser hypotheses too—hunches which they wish to
test. But both will eventually need to look for some form of
proof. In general, a historical interpretation or account becomes
credible when the pattern of evidence is consistent, and is drawn
from more than one viewpoint. Great care needs to be taken with
each of these conditions. Thus a single 'case study' is almost
inevitably a weaker base for arguing general historical interpretations than a comparison between two or more groups, each
with different characteristics, at the same period. A comparison
between different groups over time is stronger still, although
harder to achieve. The more that an argument can be shown to
hold under varying conditions, the more convincing the proof.
However, since history is made up of a multitude of cases, almost
all of which are unique in more than one way, it is in practice
often very difficult to make useful comparisons. The proof of the

explanation must then be sought from within the single case; the evidence counter-checked as far as possible in detail, and the likelihood of overall bias in it weighed. For example, in a recent study of Frontier College, the great Canadian experiment in working-class educational self-help, George Cook found himself forced to accept that he was collecting within a single broad perspective:

Generally speaking, we are hearing from those who want to help the college. Although many felt that they had 'failed' as labourer-teachers, they remain convinced that it was a 'noble idea' and reflect favourably on their experiences. They have rose-tinted glasses . . . We have not been able to reach those who have negative views . . . the early employers . . . (or) any of the early union men who worked with the college. Most importantly, we cannot find any of the labourers . . . We shall probably learn little or nothing about what *they* thought.[11]

In the same way, it would be difficult, in a study of work experience, to obtain a critical view from long-service employees who had given their lives to the enterprise, and only done so because they were prepared to accept its conditions. The upper servants of a country house provide an example. Yet while such employees are relatively easy to locate, the transient workers who may even have outnumbered them are inevitably much harder to trace. Nor, it must be strongly emphasized, will the use of written documents necessarily compensate for such an imbalance in the oral evidence. John Toland founded his sympathetic portrait of *Adolf Hitler* as 'a warped archangel', a misunderstood, 'complex and contradictory' character, on interviews with 250 survivors of Hitler's own circle.[12] He had no difficulty in buttressing it from the German archives. Oral history of this kind simply parallels the distortions of official history. It would have been a different matter had he chosen to encounter some of Hitler's opponents and victims.

A special caution is also needed if counting is to be used as part of the proof, because of the difficulties in retrospective sampling. Tabulation can be a very valuable way of classifying and disciplining one's impressions of the contents of a number of interviews. A careful scrutiny of interview material with a coding frame in mind can indeed force a much more precise consideration of what one is trying to show and what evidence the interviews can offer. On the other hand, even with interviews collected on a representative sample basis, it is best to stick to the simpler forms of analysis and not venture beyond straightforward percentages and strong correlation patterns. For example, Trevor Lummis analysed a set of thirty-five interviews for an Open University programme on 'Historical Data and the Social Sciences', concerning the decline of domestic service in the early twentieth century. It has been suggested that one reason for this decline might have been that middle-class employers wanted a more private family life, and that the presence of servants increased distance between family members. A first look at the interviews suggested, however, that the social gulfs within the household were less marked when it contained young children. Deciding to take daily eating habits as a test, he was able to produce the following tabulations:

| Household with | Households where servants ate apart from employers (%) | Households where servants shared at least one meal a day (%) |
| --- | --- | --- |
| One servant and children | 8 | 92 |
| One servant without children | 80 | 20 |
| Two servants and children | 67 | 33 |
| Two servants without children | 100 | 0 |

These figures show quite conclusively that within such households the presence of children does reduce social separation at mealtimes. They also suggest that the number of servants in the household may also be critical, but they do not prove this: it would require more figures from larger households to do so. However, providing that numbers are sufficient, and that sources of bias from the selection of informants has been allowed for, the historian can take comfort from the social scientist, for in quantitative studies the normal effect of misremembering is to lower all the correlations among variables, blurring all the patterns in a randomly confusing way rather than distorting them in particular directions. As Richard Jensen puts it, 'this means that the *true* values of correlations are higher than the observed ones. In other words, if the historian spots an interesting pattern using error-laden data, he or she can be confident that the time pattern was even stronger—certainly a happy result.'[13]

Simple counting and percentaging can be done by anyone. A pocket calculator will speed the process, but with a set of a hundred or less interviews more elaborate mechanical aids are likely to lose more time than they can save. Even with simple personal computers, you will need time to put your information on to the machine in the appropriate form; and if you are using a mainframe computer belonging to an institution, you are also likely to waste a lot of time getting the information out again, because the computer is not available just when and where you want it. The readily available programs which have so far been developed for statistical life story analysis are likely to seem too crude and cumbersome for use with transcribed interviews.[14] And the really time-consuming stage, whether or not you can use such aids, will be in the critical, detailed reading and categorization of your material.

Preliminary counting can suggest how an interpretation might be developed. But by raising new questions, it may also point to the need for further field-work. We cannot in fact make the neat separation which we have so far assumed. The ideal situation is very different: a continuous development through the to and fro of big theories, small hunches, and the practical strategy of field-work. What was initially seen as the main problem may turn out to be a misconception, a dead end; so as the field-work continues, the emphasis is shifted to another area of questioning, or a different key group of informants is sought out. Alternatively, the original theory does not fit the facts discovered. Can the theory be modified? Or is it better to look at the facts from another quite different perspective? There is, of course, no set procedure by which such a developing search for interpretation can be carried forward. By definition it demands flexibility and imagination. Not all will succeed. Scaling the historical heights is dangerous.

And few really interesting problems are ever *finally* solved. Nevertheless, in the imaginative combination of interpretation and field-work, the individual historian does have a particular advantage over the large-scale research project. Because the material can be looked at as a whole, and also in depth, from many perspectives, and because the field-work is under direct control, interpretive flexibility can be developed in a way which supports the overall objective. Indeed, the whole method is based on a *combination* of exploration and questioning in the dialogue with the informant: the researcher comes hoping to learn the unexpected as well as the expected. Hence recognized effectiveness of life story interviews in generating 'concepts, hunches and ideas, both at the local and situational level and on a historical structural level, and *within* the same field, and in relationship to other fields'.[15] By contrast it is a well-known defect of large-scale operations that although they can encompass a much wider range of possible explanations and sources, they cannot be subjected to such subtle control and modification in detail. They set out from an established research design, team work is organized on that basis, time is finite, and the field-work must be completed well before the first draft of the final report is written. Yet once the analysis of field-work is started, it becomes clear that much of the material is of little interest, but if only that particular area had been more deeply explored . . . The individual historian will not be satisfied without that further search.

One can put this in another way, by comparing the historian with a scientist. Scientific research advances through a meandering sequence of general theory, observations and hunches, experiments, working hypotheses tested by further experiments, dead ends, and further hunches and tests, until at last one hypothesis stands up to all conditions, and, if appropriate, a reformulation of theory is then sought. Any historical work suffers the inevitable disadvantage of having to work from the real cases available rather than from specially created experiments. As Edward Thompson has suggested, historians have to test their ideas with a logical process closer to that of proof in law, always vulnerable to the discovery of subsequent evidence.[16] But the big project using a field-work survey is doubly handicapped by telescoping into one all the experimental steps of the central stages of research development. It is therefore immobilized by any discovery important enough to challenge its own pre-set terms. Hence the tendency of survey findings to elaborate the obvious. They purchase their greater resources at the expense of—to use Jan Vansina's phrase—'the power of systematic doubt in historical inquiry': the very essence of creative advance in historical interpretation.

All this is somewhat abstract. Let us consider an example of the interaction between theory and field-work in practice. Peter Friedlander has set out unusually clearly, in the introduction to *The Emergence of a UAW Local 1936–1939: A Study in Class and Culture*, how his research proceeded.[17] He had at his disposal at the start certain facts—like gross census figures, dates, and a bare narrative from contemporary documents; and also various general theories—such as the Marxism of class struggle underlying labour history, and from Max Weber the concepts of rationality and individualism as essential to a bourgeois epoch. But the gaps were enormous. There was no documentary evidence of attitudes in the factory to authority and how this changed as the trade union was organized; of who made up the inner circle of union leaders and how they were related to social groups within the factory and whether these leaders led or reflected opinion; or of which were in fact the key social groups of workers in the factory, how their attitudes to the union struggle varied, and how it affected their personal lives and outlooks. Equally, the theoretical concepts failed to meet. This trade union struggle took place not merely within a highly developed industrial capitalist society. The majority of the workers had migrated into the city where they worked from quite different social contexts. Their fight to unionize was thus also part of a much wider transformation of social cultures in migrant families and individuals: in this case religious-minded Slavs, revolutionary Croatian nationalists, Yankee and Scots artisans, Appalachian farm families and urbanized American Blacks. These specific cultural sub-groups were in the event to provide the key to interpretation. Yet, as Friedlander observes, labour historiography, which has tended to assume the presence of a modern, individuated, rational worker, has usually viewed the process of unionization in narrowly rational, institutional, and goal-orientated terms. The problem of culture and praxis is passed over in silence.

Even where an explicitly Marxist framework is used in labour history, the tendency is for a whole section of society to be

conceived of as an individual, and the problem is then to explain the institutional formation as the outcome of a rational process within the consciousness of this *quasi* individual.

But it is not always easy to locate this expected rationality; nor to explain its shortfall *in a particular case* in terms of general theoretical concepts such as, for example, 'false consciousness':

At each juncture where a gap is seen between the abstractions of the political economy of work, and the concrete reality of individual, peer group, gang, clique, family, and neighbourhood—of character and culture—there appear *ad hoc* psychological notions invested with an astonishingly ubiquitous explanatory power. Such notions ignore one of the basic problems of historical thought: the nature of relationships among these many layers of social reality . . . the complex structure of cultures and relationships that develop and interact.

As the research proceeded, it emerged that only the older-established American skilled Protestant workers could be described in classic individualistic and rationalistic terms. This group supplied most of the leadership, although it also included many who felt no interest in the union. The Appalachians also acted as individuals, but principally on a moral basis: they joined the union relatively late, when they believed that its cause was *right*, and once having joined were as utterly loyal as to their religious sects. The older East European migrants were much more concerned with what was right or wrong in social or ethical terms for the community, and acted explicitly as a group. Although personally cowed and submissive, they disliked the foremen and the management, and became dependable supporters of the union leadership. Their children, by contrast, were much more active and outspoken, and in particular a group of young Poles who belonged to neighbourhood gangs played a special role in the struggle. Like the older Slavs, they acted together, but with little social and political consciousness: they were pragmatic, opportunistic—the uncontrollable militant wildcatters willing to break a contract by striking, and then to man the flying picket squad. It was as if the union to them was 'a bigger and better gang'.

It was only when these groups and their attitudes had been identified, that the narrative of the struggle could be meaningfully reconstructed. Yet not only was none of this information available at the start, but it was not even known to be needed. The discovery of information and development of an interpretation went forward hand-in-hand as, over a period of eighteen months, Friedlander talked with the union leader, Edmund Kord. Kord had an exceptionally full and accurate memory, and indeed, remembered more as his mind became increasingly focused on these past years. Friedlander spent a full week with him three times, and each of these prolonged sessions produced drafts, comments, questions, and discussion. One of the two intervals between sessions included six hours of recorded telephone discussion; another produced altogether seventy-five pages of correspondence. They had to create between them not just the facts which were needed, but a mutual understanding and language of exchange. And if the 'thick description' into which Friedlander finally fuses both facts and interpretation does not allow him the last clear step into a new theory, he certainly laid the grounds for it in the marked differences which he shows between generations as well as between various social groups in the factory, in their particular roads from one consciousness to another.

The contrasting paths taken by different generations of the same workgroup are also demonstrated by Tamara Hareven's remarkable studies of Manchester, once the textile capital of New England. Founded by the Amoskeag Company in the 1830s, the city grew around its booming millyard and the promise of steady, well-paid work drew in successive waves of immigrants. By the early twentieth century its complex of thirty mills, employing 17,000 workers, made up the largest textile plant in the world. The giant works was so central to their lives that the people of Manchester believed it would stand for ever: 'You thought it would always be there.' Yet within two decades, undercut by cheaper labour and newer machinery in other regions, the giant was dead. Amoskeag closed its doors, bankrupt, in 1936. Smaller firms later revived parts of the millyard so that textile work struggled on for another forty years in Manchester, but the last mill finally shut in 1975. Even then, there were workers who left in tears: 'I'll miss the people I worked with, I'll miss the mill itself . . .'; 'it's like a second home'.[18] The industrial revolution had come and gone: a haunting allegory of the fate of much of the Western world.

Tamara Hareven has published two books about Manchester. The first, *Amoskeag* (1978), was a dramatic documentary built around Randolph Langenbach's photographs and the testimonies of former mill workers: about getting jobs and learning skills, the pleasures and tensions of work, larking about, company paternalism, and the bitter dying struggles with Amoskeag. It is a testament to industrial work, its centrality to people's lives, and the jeopardy in which that work now stands, told through the men and women of Manchester themselves—a book of rare power. *Family Time and Industrial Time* (1982), by contrast, is a reflective and analytical interpretation marshalling a much broader range of source material. Alongside extracts from the interviews, the arguments are backed by numerous tables from the local census and from a sample of the Amoskeag's workforce records. Hareven provides a more fully documented labour history of the Amoskeag's evolving policies of paternalism, scientific management, confrontation with labour, and company unionism, as well as analyses of career patterns and opportunities for promotion within the mills.

The most important insights of the book come however from the juxtaposition of this study of the factory world with the family lives of Manchester workers which is made possible through oral history. The result is to challenge many widely held views. She shows how it is not the 'modern' nuclear family which deals best with a catastrophe on the scale of widespread redundancy, but the more 'traditional' extended family, which can remain effective when scattered—indeed, *more* effective just because it is scattered. The extended family had been the channel of recruitment of migrant workers to the mill; and at the end, it was the safety net of the retreat. Or again, workers who had *not* had steady careers proved more likely to have the adaptability to face such a crisis successfully than those who had. Such findings are set, moreover, within a clearly articulated theoretical frame of 'family time' and 'industrial time': the cross-cutting struggle of family 'life plans' and industrial history. The clock analogy perhaps suggests too much certainty in the outcome, but it brings out well how, though some aspects of the life cycle were constantly repeated, the experience and the chances of each generation differed sharply. While to one the Amoskeag gave the security of a paternalistic family, and chances of promotion, to the next generation it offered a nightmare of tension; and to the last the hopelessness of a sinking ship. The twisting consciousness of the community—loyal, militant, despairing—reflected the historic moment at which the youth of each generation entered the factory gate.

This ability to make connections between separated spheres of life is an intrinsic strength of oral history in the development of historical interpretation. In studying the transition from one culture to another, in time, or through migration, we can not only look at those cultures separately, but observe the paths that individuals took from one culture to another. And almost every individual life breaks across the boundaries between home and work. Escaping from these conceptual boxes can produce strikingly new hypotheses even from a small-scale study. In demography, for instance, it has been assumed that family limitation and the use of birth control spread by the 'diffusion' of attitudes from the professional middle classes down the social scale to the working classes. Some exceptions, like the low fertility of cotton workers, had been noted, but it was a pilot oral history project by Diana Gittins which first indicated that the basic 'diffusion' model was false: for working-class women changed their birth-control practices through independent influences—notably discussion at work—rather than direct middle-class influence. Indeed, those with the closest contact with middle-class families, who worked for them as domestic servants, received the least advice on family limitation; and even doctors and nurses were generally unhelpful, if not positively misleading, to working-class patients. This first exploratory discovery through oral history led to the substantial research, including statistical analyses of women workers' fertility rates and the use of early clinic records, which Diana Gittins has published in *Fair Sex* (1982). Her reinterpretation is a typical outcome of oral history, for 'diffusion' theory gives credit to the middle classes for a social transformation which owes as much to the aspirations of working-class women themselves.

Yet if working women have played such a crucial part in the profound social change marked by the demographic transition from the 1870s to the 1920s, from which so much else, economic and social, has followed, why have they been so much slower than men in recognizing their collective self-interest in politics and trade unionism? Male politicians and historians have too often assumed it 'natural' for women to take a less active part in the labour movement; and when the problem has been considered at all, it has been in terms of the workplace, and women's shorter, more interrupted working lives. But Joanna Bornat's research on Yorkshire textile unions has shown how women's consciousness was shaped as much through subordination at home as in the factory. They found their jobs through family contacts, were trained by kin at the mill, and handed back their entire wage packets to their mothers; and it was their fathers who decided whether or not they should join the union. If they joined, collectors took their subscription at the doorstep, not on the mill floor.[19] In short, the male division of the worlds of work and home has obscured any adequate understanding of the class-consciousness of women workers. But a history which cannot account for them rests on flawed foundations.

It is undoubtedly a danger that oral sources, used on their own, can encourage the illusion of an everyday past in which both the cut and thrust of contemporary political narrative and the unseen pressures of economic and structural change are forgotten, just because they rarely impinge directly on the memories of ordinary men and women. It is essential to place them in this broader context. But as we have seen, oral sources can also help us to understand how that context is itself constituted. They offer the promise, moreover, of advancing this understanding in a fundamental way.

They suggest, first, a basic misconception in the dynamics of social change. These are almost always described in terms which reflect the experience of men: of collective and institutional rather than personal pressures, of the logic of abstract ideology, acting through the economy, through politics, through élite networks of unions and pressure groups. Behind are the deeper contradictions of social and economic organization which sometimes openly, sometimes unknowingly they express. But an equally crucial element is missing: the cumulative effect of individual pressure for change. It is this which immediately emerges through life histories: the decisions which individuals make—to move or improve a house, to leave one community and migrate to another; to leave a job which has become intolerable or to look for a better one; to put money into the bank, or shares, or a business of one's own; to marry or to separate, to have or not to have children. The changing patterns of millions of conscious decisions of this kind are of as much, probably more, importance for social change than the acts of politicians which are the usual stuff of history.

This becomes evident as soon as we look at the major long-term social changes of the Western world in the last century. Certainly the ebb and flow of political rights and civil liberties, and the growing state intervention in education and welfare, have been the outcome of collective pressure and political decision; and collective trade union pressure has kept up the working-class share in real earnings and cut the hours given to paid work. But this does not touch the two most startling changes: the rise in economic productivity and living standards, and the reduction in the number of children. Neither are the result of political intervention—indeed, no state has yet shown much ability to influence either, except through inducing unintended disaster. The truth is that the mechanics of change of both the economy and population, although basic to everything else, are very imperfectly understood.

They will remain so until we incorporate, as part of the structure of interpretation, the cumulative role of the individual. That implies recognizing that a high proportion of crucial individual decisions are as likely to be made by women as by men—not only in spheres like family-building, but also as migrants and as workers (women change jobs more frequently than men). Equally important, we need to know how public ideas, economic and collective pressures interact at an individual level—as in the seizing of economic chances, or in the shaping of attitudes through family and friendship and the media, and through childhood and adult personal experience—to form those myriad decisions which cumulatively not only give shape to each life story but can also constitute the direction and scale of major social change. Or to put it another way, it becomes clear that the production of people is as much the powerhouse of change as the production of things.

An example may again help. When I began the research for *Living the Fishing* (1983) I assumed that the economy would shape family relationships, and it indeed proved true that women in fishing families in many parts of the world, because of the frequent absence of their men at sea, take a greater share of the responsibility and authority in the family; although this may range from the 'partnership' marriage common among inshore fishermen whose wives work with them in a joint enterprise gutting and marketing the fish, to the long-distance, deep-sea fishermen who are effectively absentee husbands leaving their wives as single parents. Untangling the variations in between revealed a complex of other influences too, in which economy, property, space, work, religion, and family culture all played a part.[20] But economic influence did not work in one direction only. In a wage-earning company port like Aberdeen life aboard became so rough and family life so battered by drink and violence that the next generation voted with its feet; mothers sent their sons to look for other work, and young women looked for other husbands than fishermen. Family culture was equally critical to the economic survival of family boat-owning communities, but in a very different way. Here the widespread encouragement of individual initiative among the fishermen was needed to ensure recurrent adaptability in the face of rapidly changing fish stocks, technology, and markets. Part of the secret of the most successful ports turned out to be the inculcation of an ideology of hard work, thrift, achievement, and independence from childhood. But this valuing of individual worth had to go with an acceptance of some eccentricity as the price of creativity. And the transmission of such values was encouraged by the affectionate gentleness typical of a Shetland upbringing, where children were encouraged to talk and reason for themselves in a relatively egalitarian home; while it was severely inhibited by the more authoritarian, punitive, hierarchical, male-dominated family characteristic of Lewis. With apparently equal chances, the fishing of one flourished, while the other withered.

Certainly the constraints exercised by the economic system, technology, and resources on how men and women live their lives are fundamental. But the economy is a social creation, and part of its making is in the family. The unpaid labour of women within the household not merely services it, but also, through the rearing of children, the workforce of the future, lays part of the foundations of the future. Clearly both the transmission of values between generations and the moulding of personality within the family are questions of critical importance to historical understanding. They demand examination at many different levels, including, as we have seen earlier, that of cultural patterns and emotional configurations repeating themselves over the generations in different families.[21] But bringing these together will also require a major imaginative leap in our use of theory.

At present we can turn to one of two general types of theoretical interpretation. On the one hand there are the big theories of social organization, social control, the division of labour, the class struggle, and social change: the functionalist and other schools of sociology and the historical theory of Marxism. On the other hand there is the theory of individual personality, of language and the subconscious, represented by the psychoanalytical approach. They can be layered together, as in an individual biography, but no satisfactory way has yet been found of bonding them. Psycho-history has simply resorted to the crude device of 'analysing' whole groups—even whole societies—as if they were a single individual with only one life experience. The difficulties in any more subtle reconciliation have emerged very clearly in the debates on Marxism, feminism, and women's history. The fundamental problem lies in the fact that each type of theory turns its back on the other. Marxism, like sociological theory in general, is deliberately concerned with minimizing the role of the individual, as opposed to the social group. Psychoanalysis claims to be founded upon the elemental human personality, and so independent of history. Yet while Marxism rests on the belief that men and women create their consciousness through what they do, the archetypal Freudian psychoanalysis assumes that the fundamental shaping of personality is completed in infancy—before the limits of remembered conscious action. This leaves few clues as to how a bridge between the two types of theory can best be constructed. It is nevertheless an essential task if history is to provide a meaningful interpretation of common life experience. And in this task, oral history will have a vital role. Its evidence intrinsically combines the objective with the subjective, and leads us between the public and private worlds.

It is only by tracing individual life stories that connections can be documented between the general system of economic, class, sex, and age structure at one end, and the development of personal character at the other, through the mediating influences of parents, brothers and sisters, and the wider family, of peer groups and neighbours, school and religion, newspapers and the media, art and culture. Only when the precise role of these intermediary institutions in, for example, socialization into sex and class roles, has been established, will a theoretical integration become a possibility. Until then we can only guess how far the economic and social system moulds personality, or the system is itself shaped by basic biological drives. A beginning to such work can be seen but it would be foolish to claim more than this as yet. It represents, nevertheless, for the future probably the greatest challenge and contribution which oral evidence may offer to the making of history.

Ten years ago I ended this book with a brief look forward into that future. Much of what I then hoped for has come about. Oral history has been vindicated in a whole series of publications, both empirical and theoretical. While the most hidebound opponents continue to snarl, mostly in private, the main debate has shifted from whether to use oral history or not, to how best to use it. It has left a more general awareness of how all historical evidence is moulded by individual perception and, selected through social bias, conveys messages of prejudice and power. The nature of history in this double sense has been an issue too long evaded by historians.

More than that, entirely new uses of history have sprung up in the movements for reminiscence therapy and reminiscence drama. More generally, there has been a quickening change in resources. Local and regional oral history collections in public libraries and archives have spread rapidly, and national sound archives are following. It already requires less unusual imagination and effort for a teacher to use recordings, or a museum to incorporate them in a display. In time it will become relatively easy to find an extract published on tape of a particular person, event, or theme in either political or social history.

Unique, often disarmingly simple, epigrammatic, yet at the same time representative, the voice can as no other means bring the past into the present. And its use changes not only the texture of history, but its content. It shifts the focus from laws, statistics, administrators, and governments, to people. The balance is altered: politics and economics can now be seen—and thus judged—from the receiving end, as well as from above. And it becomes possible to answer previously closed questions: extending established fields such as political history, intellectual history, economic and social history; adding to other newer areas of inquiry—working-class history, women's history, family history, the history of racial and other minorities, the history of the poor and of the illiterate—a whole new dimension. We have already in existing titles—*Akenfield*, *Where Beards Wag All*; *Working, Workless*; *Pit-men, Preachers and Politics*, *From Mouths of Men*; *Division Street, The Classic Slum*; *Below Stairs, The Children of Sanchez*; *All God's Dangers, Blood of Spain*; *The Dillen, The Leaping Hare*—the first swallows of a new summer. As others follow, history will be changed and enriched.

The new balance to the content of history, and the sources of its evidence, will alter its judgement, and so, eventually, its message as public myth. We shall find from the past a different set of heroes: ordinary people as well as leaders; women as well as men; Black as well as White. History, which once could only weep for a King Charles I on the scaffold, can now share grief with the old illiterate widower, Nate Shaw, twice-arrested black Alabama sharecropper, at the loss of his wife Hannah:

I just felt like my very heart was gone. I'd stayed with her forty-odd years, and that was short, short—except bein pulled off and put in prison. I picked her out amongst the girls in this country and it was the easiest thing in the world to do . . . She was a Christian girl when I married her. And she was a woman that wanted to keep as far as her hands and arms could reach, all the surroundings, she wanted to keep it clean. And I've kept myself clean as I possibly could. But in past days, I've sneaked about in places, I did. I own to my part of wrongness . . . I liked women, but . . . I desperately kept clean of runnin too much to a extreme at other women when I had her. Regardless of all circumstances, I weren't a man to slip around at women and no matter what I said to another woman or what I done, I let my wife come first . . . I'm praisin her now, I'm praisin her for what she was—she was a mother for her children, she was a mother for her children—and when they put me in prison, the whole twelve years, she stayed by her children, she didn't waver . . . I loved that gal and she dearly proved she loved me. She stuck right to me every day of her life and done a woman's duty. Weren't a lazy bone in her body and she was strict to herself and truthful to me. Every step she took, to my knowledge, was in my favour. There's a old word that a man don't ever miss his water until his well go dry . . .[22]

There will be more biographies like Nate Shaw's. Whose, we can only guess. A London West Indian bus conductor; a British Leyland assembly-line car worker; a Belfast boiler-maker's wife; a supermarket checker-out; a Welsh sheep-farmer; a Pittsburgh steelmaker; a Californian telephonist; a New South Wales truck-driver . . . Who knows? Or what particular questions oral history will succeed in solving. The riddle of British working-class Toryism? Whether the old family firm was an economic asset or a handicap? How far industrialization emancipated women, or confined them as housewives to still more limiting male domination? What makes some social groups prefer to educate, and others to beat their children? How some persecuted immigrant minorities prosper, and others not? In what social context are major scientific discoveries made? To each of these problems, oral history could make a critical contribution. Which are chosen depends on who sees this first.

In principle, the possibilities of oral history extend into every historical field. But they are more fundamental to some than to others. And they provide an underlying current: towards a history which is more personal, more social, and more democratic. This affects not only published history, but the process by which it is written. The historian is brought into touch with fellow-scholars in other disciplines: social anthropology, dialect and literature, political science. The academic is prised out of the closet into the outside world. The hierarchy of higher and lower institutions, of teachers and taught, breaks down in joint research. Old and young are brought into exchange and closer sympathy. The classics of oral history will no doubt continue to be created by uncategorizable individuals. But there has been a quiet sea-change in the process of historical writing, scarcely noticed by the book reviewers. Increasingly small oral history groups have been bringing out their own publications. Certainly most would gain from more interpretation, and often only a local could make the most of all the detail. It may be a history of the street and its families; of a factory's workers and owner; about a strike, or a bomb explosion; recollections of past leisure, education, or domestic service. These local publications are gathering new historical material for the future, which would have been otherwise lost. They are tapping the river water at the sea's mouth. The far limit of the past recoverable through oral evidence recedes remorselessly through death, day by day. But the real justification of history is not in giving an immortality to a few of the old. It is part of the way in which the living understand their place and part in the world. Landmarks, landscapes, patterns of authority and of conflict have all been found fragile in the twentieth century. By helping to show how their own stories fit into the changing character of the place in which they live, their problems as workers or as parents, history can help people to see how they stand, and where they should go. This is what lies behind the present popularity of recent history in Britain. It also points to the key social and political importance of oral history. It provides a new basis for original projects, not just by professionals, but by students, by schoolchildren, or by the people of a community. They do not just have to learn their own history; they can write it. Oral history gives history back to the people in their own words. And in giving a past, it also helps them towards a future of their own making.

# Further Reading and Notes

*Abbreviations: Oral History—OH; International Journal of Oral History—IJOH; Life Stories—LS; Oral History Review—OHR; Canadian Oral History Association—COHA; Bulletin of the Society for the Study of Labour History—BSSLH; History Workshop—HW.*

Place of publication is only given for works which were not published in London.

These suggestions for further reading and notes follow the subject pattern of the chapters.

For a general introduction, two classic books stand out: Jan Vansina, *Oral Tradition: a study in historical methodology*, 1965 (modified reissue, *Oral Tradition as History*, 1985), and George Ewart Evans, *Where Beards Wag All: The Relevance of Oral Tradition*, 1970—the first based on historical field-work in Africa, the second in East Anglia. More recently, Ken Plummer, *Documents of Life: An Introduction to the Problems and Literature of a Humanistic Method*, 1983, has provided an excellent introduction for social scientists. David Henige, *Oral Historiography*, 1982, provides an up-to-date overview on the collecting of oral tradition in the Third World.

However, the most rewarding sources for the continuing development of oral history are journals: notably *Oral History, the journal of the Oral History Society* (Department of Sociology, Essex University, Colchester, England); the *International Journal of Oral History* (Meckler Publishing, 11 Ferry Lane West, CT 06880, USA), and the *Oral History Review* and *Newsletter* of the Oral History Association, USA (PO Box 926, University Station, Lexington, Kentucky 40506, USA). There are national journals in Canada, Australia, Italy, and other countries too, as well as many local journals: the best known of all is the school magazine *Foxfire* (Rabun Gap, Georgia 30568). For Britain and the United States directories of archives and projects are also available, from the Oral History Society and the Oral History Association respectively.

Finally, three notable collections bring together articles and conference papers: Daniel Bertaux (ed.), *Biography and Society: The Life History Approach in the Social Sciences*, 1981; David K. Dunaway and

Willa K. Baum (eds.), *Oral History: An Interdisciplinary Anthology*, 1984; and Paul Thompson (ed.), *Our Common History: The Transformation of Europe*, 1982 (papers from the 1979 international oral history conference).

Introductions available in other languages include Lutz Niethammer (ed.), *Lebenserfahrung und Kollektives Gedächtnis die Praxis der 'Oral History' Herausgegeben*, Frankfurt, 1980; Martin Kohli and G. Robert (eds.), *Biographie um Soziale Wirklichkeit, neue Beitrage und Forschungsperspektiven*, Stuttgart, 1984; Philippe Joutard, *Ces voix qui nous viennent du passé*, Paris, 1983; B. Bernardi, C. Poni, and A. Triulzi (eds.), *Fonti Orali: Antropologia e Storia*, Milan, 1978 (papers from the Bologna 1976 conference, in Italian, English, and French); Luisa Passerini (ed.), *Storia Orale: vita quotidiana e cultural materiale delle classi subalterne*, Turin, 1978; and Franco Ferrarotti, *Storia e storie di vita*, Rome, 1981. Special oral and life-history numbers have been published in French by *Annales* (30, 1, 1980) and *Cahiers internationaux de sociologie* (LXIX, 1980).

## Chapter 1   History and the Community

For recent British developments, see *Oral History and Community History*, special number, *OH*, 12, 2, and Rickie Burman, 'Participating in the Past? Oral History and Community History in the Work of Manchester Studies', *IJOH*, 5, 2. On the social purposes and manipulation of history: David Lowenthal, *The Past is a Foreign Country*, Cambridge, 1985; Eric Hobsbawm and Terence Ranger (eds.), *The Invention of Tradition*, Cambridge, 1983; Jean Chesneaux, *Pasts and futures, what is History for?*, 1978.
1. *OH*, 1, 2, p. 9.
2. *OH*, 1, 4, p. 57.

## Chapter 2   Historians and Oral History

For oral history in non-literate societies, see Vansina, *Oral Tradition*; Henige, *Oral Historiography*; D. F. McCall, *Africa in Time Perspective*, 1964; Joseph C. Miller (ed.), *The African Past Speaks: essays on oral tradition and history*, Folkestone, 1980; Walter Ong, *Orality and Literacy*, 1982; John Miles Foley (ed.), *Oral Traditional Literature*, Columbus, 1981; D. L. Page, *History and the Homeric Iliad*, Berkeley, 1959. For the changing character of perception in the first age of printing: Robert Mandrou, *Introduction to Modern France 1500–1640, An Essay in Historical Psychology*, 1975. A sustained critique of both oral and printed sources over three centuries on a single topic, the Protestant defiance of the French state in the Cévennes, is in Philippe Joutard, *La Légende des Camisards: une sensibilité au passé*, Paris, 1977. On historians: J.W. Thompson, *A History of Historical Writing*, New York, 1942.

For the subsequent development of local history, see Raphael Samuel, 'Oral History and Local History', *History Workshop*, 1. For the early development of social survey methods in Britain and Europe, see Anthony Oberschall, *Empirical Social Research in Germany 1848–1914*, 1965; and Marie Jahoda, Paul Lazarsfeld, and Hans Zeisel, *Marienthal*, trans. 1972, p. 100 ff. For autobiography in England, see David Vincent, *Bread, Knowledge and Freedom: a study of nineteenth-century working-class autobiography*, 1981, and for the earlier religious autobiographies, Owen C. Watkins, *The Puritan Experience: studies in spiritual autobiography*, New York, 1972; in Germany, Oberschall; in France, Philippe Lejeune, *Je est un autre: L'autobiographie, de la littérature aux médias*, Paris, 1980.

Full accounts of related developments in folklore are provided by R.M. Dorson (ed.), *Folklore Research Around the World*, Bloomington, 1961; Folke Hedblom, 'Methods and Organization of Dialect and Folklore Research in Sweden'; Sean O'Sullivan, 'The Work of the Irish Folklore Commission'; Alan Bruford, 'The Archive of the School of Scottish Studies'; Stanley Ellis, 'An Introduction to the Work of the (Leeds) Institute of Dialect and Folk Life Studies', *OH*, 2, 2 and 4, 1; and Roger Abrahams, 'Story and History: a folklorist's view', *OHR*, 1981. For oral history, history, and sociology, see the review article, Martin Bulmer, 'Sociology and History: Some Recent Trends', *Sociology*, 8, 1, pp. 137–50, and Philip Abrams, *Historical Sociology*, 1982, ch. 9.

On the origins of French oral history: Joutard, *Ces voix*; for fuller surveys of the oral history movement in various regions: Paul Thompson, 'Oral History in North America'; Rolf Schuursma, 'The Sound Archive of the Film and Science Foundation and the Dutch Radio Organisation'; Paul Thompson, 'The Bologna Conference' and 'Oral History in Israel'; Andrew Roberts, 'The Use of Oral Sources for African History'; Terence Ranger, 'Personal Reminiscence and the Experience of the People in East Central Africa'; Paul Thompson, 'The New Oral History in France'; Jorgen Burchardt and Carl Erick Andresen, 'Oral History, People's History and Social Change in Scandinavia'; Paul Thompson, 'Oral History in Belgium'; Stephen Thompson and Yang Li-wen, 'Oral History in China'—*OH*, 3, 1; 1, 2; 5, 1; 4, 1; 6, 1; 8,

1; 8, 2; 11, 1; and 15, 1, respectively; Aspasia Camargo, Valentina da Rocha Lima and Lucia Hippolito, 'The life history approach in Latin America', *LS*, 1985; Anna Bravo and David Ellwood, 'Oral history and resistance history in Italy', Sven Lindqvist, 'Dig where you stand' and Paul Thompson, 'The humanistic tradition and life histories in Poland', in *Our Common History*.

1. Michelet, *Histoire de la Révolution Française*, Paris, 1847, 2, p. 530: 'la tradition orale'; Thompson, 2, p. 241.

2. Reissued as *Oral Tradition as History*, 1986, expounding a similar viewpoint, although more cautiously.

3. D. T. Niane (ed.), *Sundiata: an epic of old Mali*, 1965, p. 1.

4. Haley, 'Black History, Oral History and Genealogy', *OHR*, 1973, pp. 14–17.

5. Trans. L. Shirley-Price, 1955, p. 34. On the Spanish friars in Mexico, Georges Baudot, *Utopie et historie au Mexiave*, Paris, 1977.

6. Thus of 164 extant Edward the Confessor charters, only 64 are accepted as authentic: M. T. Clanchy, *From Memory to Written Record: England 1066–1307*, Folkestone, 1979, p. 249.

7. p. 89.

8. *Works,* trans. W. F. Fleming, New York, 1927: V, p. 62, XI, p. 9 and XVIII, pp. 6, 8 and 15; Thompson, 2, p. 67.

9. James Boswell, *Journal of a Tour to the Hebrides with Samuel Johnson*, Ll. D, 1785, pp. 425–6.

10. Trans. R. Rawlinson, 1728, pp. 276–8.

11. Thompson, 2, p. 67.

12. I, pp. 382–4, 418.

13. David Vincent, 'The decline of oral tradition in popular culture', in R. D. Storch (ed.), *Popular Culture and Custom in Nineteenth-century England*, 1982; Marilyn Butler, personal information.

14. *From Mouths of Men*, 1976, p. 179; Sir Walter Scott, *Tales of My Landlord*, 1816.

15. Oliver Lawson Dick (ed.), *Aubrey's Brief Lives*, 1949, p. xxix.

16. 1968, p. 1 (Hoskins); David G. Hey, *An English Rural Community: Myddle under the Tudors and Stuarts*, Leicester, 1974.

17. James Everett, *Wesleyan Methodism in Manchester and Its Vicinity*, 1, 1827; and the London Baptist minister Josiah Thompson was compiling testimonies on early dissenting congregations in 1772–84, now in Dr Williams Library, London (information from John Walsh); Antoine Court among French Protestants in the 1730s.

18. Smith, *The Working Man's Way in the World, being the Autobio-*

*graphy of a journeyman printer*, 1853; also, Eleanor Eden (ed.), *The Autobiography of a Working Man*, 1862—especially remarkable for its lively, almost spoken style. The *How I Became a Socialist* series of the 1890s illustrates the conversion testimonial in reverse. See the full listings, including over a thousand nineteenth-century authors, in John Burnett, David Vincent, and David Mayall (eds.), *The Autobiography of the Working Class: An Annotated Critical Bibliography*, I–III, 1750–1945, 1985–7.

19. Oberschall, p. 81. Göhre's German series started with Carl Fischer, *Denkwürdigkeiten und Erinnergungen eines arbeiters*, Leipzig, 1904—a brick-factory worker. One of the classics of the new genre was subsequently translated into English: Adelhaid Popp, *The Autobiography of a Working Woman*, 1912. In France working-class autobiography began with militants and the 1848 Revolution and the Paris Commune of 1871, but remained a rarity; instead, working-class experience was presented in novels and newspapers, as dreams (of peasants) and nightmares (of workers): Lejeune, *Je est un autre*, p. 251.

20. 1797, p. ii.

21. Eileen Yeo, 'Mayhew as a Social Investigator', in E. P. Thompson and E. Yeo (eds.), *The Unknown Mayhew*, 1971, pp. 54–63; Henry Mayhew, *London Labour and the London Poor*, 1851.

22. Mary Higgs, *Glimpse into the Abyss*; Jack London, *The People of the Abyss*, 1903; George Orwell, *Down and Out in Paris and London*, 1933; Sherard, pp. 41–3; Harold Wright (ed.), *Letters of Stephen Reynolds*, 1923, p. 109, to Tom Woolley, 25 October 1908, and *Daily News*, 22 May 1923. See also the perceptive preface to *Seems So!*

23. *Our Partnership*, 1948, pp. 27 and 158; Margaret Cole, *Beatrice Webb*, 1945, p. 59.

24. *Economic Journal*, xvi, p. 522.

25. Michael Twaddle, 'On Ganda historiography', and Robin Law, 'Early Yoruba historiography', *History in Africa*, 1 (1974), pp. 85–100, and 3 (1976), pp. 69–89; Charles Morrissey, 'Why call it "Oral History"?', *OHR*, 1980, pp. 29 ff.; *A Brief Account of the Literary Undertakings of Hubert Howe Bancroft*, 1883; J.W. Caughley, *Hubert Howe Bancroft, Historian of the West*, Berkeley, 1946; Willa Baum, 'Oral History: a revived tradition at the Bancroft Library', *Pacific North West Quarterly*, April 1967, pp. 57–64.

26. pp. v, 3, 520.

27. pp. iv–vi, 1, 5 and 11. David Hume and Edward Gibbon worked with similar care.

28. H.P. Rickman (ed.), *Meaning in History*, 1961, pp. 85–6.

29. Trans. G.G. Berry, p. 17.

30. Acton's letter to contributors: Fritz Stern, *The Varieties of History*, New York, 1956, p. 247.

31. Langlois and Seignobos, pp. 129, 134, 155, 175, 196; Collingwood, p. 131.

32. *The Struggle for Mastery in Europe, 1848–1918*, Oxford, 1954, pp. 569–72.

33. *Listener*, 1 February 1973, p. 148; *English History 1914–45*, Oxford, 1945, p. 609; *Struggle for Mastery*, p. 574.

34. Walter Lowe Clay, *The Prison Chaplain: a Memoir of the Rev. John Clay*, Cambridge, 1861. For perceptive discussions of Chicago life story sociology, see especially James Bennett, *Oral History and Delinquency: the Rhetoric of Criminology*, Chicago, 1982, and Abrams, *Historical Sociology*, 1982.

35. H.L. Beales and R.S. Lambert, 1934.

36. B. Malinowski, 'Myth in Primitive Psychology', in W.R. Dawson (ed.), *The Frazer Lectures*, 1932, p. 97; cf. *Sex and Repression in Savage Society*, 1927, p. 104; Radin, p. viii. For the American anthropological tradition, see Plummer, *Documents of Life*; Gordon W. Allport, *The Use of Personal Documents in Psychological Science*, New York, 1942; L. Gottshalk, C. Kluckhohn, and R. Angell, *The Use of Personal Documents in History, Anthropology, and Sociology*, New York, 1947. Outstanding examples include Walter Dyk, *Son of Old Man Hat: a Navaho autobiography*, New York, 1938; Sidney Mintz, *Worker in the Cane: a Puerto life history*, New Haven, 1960; Oscar Lewis, *Pedro Martinez: a Mexican peasant and his family*, 1964; Helen Hughes, *The Fantastic Lodge*, 1961. Robert Redfield edited 76 interviews by Manuel Gamio in a thematic collection, *The Life Story of the Mexican Immigrant*, 1931. The rare British contributions to the method include the more interpretative use of life stories in Geoffrey Gorer, *Himalayan Village: an account of the Lepchas of Sikkim*, 1938; and an outstanding early woman's life story in Mary F. Smith, *Baba of Karo: a woman of the Muslim Hausa*, 1954.

37. pp. ix–x. *Somebody* was also playing with such ideas in New York: witness the caricature of Professor Sea Gull, with his vast project of 'An Oral History of Our Time', an informal history of 'the shirt-sleeved multitude', incomplete after 26 years barroom tippling and

flophouse dossing, of Joseph Mitchell's *McSorley's Wonderful Saloon*, 1938, pp. 68–86.

38. *Tianjin Daily*, 28 April and 20 June 1958.
39. Franco Ferrarotti, *Vita di baracati*, Naples, 1974 and *Vite di periferia*, Milan, 1981; Alessandro Portelli, *Biografia di una Città*, Turin, 1985; Luisa Passerini, *Torino operaia e fascismo*, Rome, 1984; Nuto Revelli, *Il mondo dei vinti* and *L'anello forte*, Turin, 1977 and 1985.
40. Lutz Niethammer, *'Die Jahre weiss man nicht . . .': Lebensgeschichte und Sozialkultur im Ruhrgebiet 1930 bis 1960*; *'Hinterher merkt man . . .'*; and *'Wir Kriegen jetzt andere Zeiten'*, Bonn, 1983–5; Maurice Halbwachs, *Les Cadres sociaux de la mémoire*, Paris, 1925. In France the study of local folklore developed in the early nineteenth century, especially in the peripheral regions of Brittany and Provence; the systematic national study of dialect and peasant ethnology was established from the 1870s, and the Musée des Arts et Traditions Populaires in Paris founded in 1937. Popular interest in oral history was awakened by Alain Prévost's *Grenadou: paysan français*, Paris, 1966, a life story of family life, farm work, and war in the northern countryside near Chartres, derived from tape recordings but thoroughly rewritten. It was televised, and inspired a series of popular autobiographies in the 1970s, including the best-selling Pierre Jakez Hélias, *Le cheval d'orgueil*, Paris, 1975 (from Brittany) and Serge Grafteaux, *Mémé Santerre*, Paris, 1975 (about a northern miner's daughter). The television showing of the film 'Le Chagrin et le Pitié' in 1971, a documentary of collaboration in a town during the German occupation based on retrospective interviews, was also an important influence in the start of oral history work.
41. pp. 12–13.
42. p. 214.
43. *SSRC Newsletter*, 31 July 1976, p. 6.
44. *From Mouths of Men*, p. 187.
45. For a recent, more moderate, but still uncomprehending critique of oral history from this standpoint, see Louise Tilly and the ensuing debate on 'People's History and Social Science History', *IJOH*, 6, 1.
46. *OH*, 1, 3, p. 46.
47. p. 142: Marwick has indeed shifted since, but very grudgingly— 1981 edition, p. 141.
48. For a bad-tempered attack on the idea that it is worth talking to

ordinary people, see the Australian history professor P. O'Farrell in *Quadrant*, November 1979, pp. 4–8, reviewing this book.

## Chapter 3  The Achievement of Oral History

Economic history: Elizabeth Roberts, 'Working-class standards of living in Barrow and Lancaster, 1890–1914, *Economic History Review*, 1977; Christopher Storm-Clark, 'The Miners, 1870–1970: a Test-Case for Oral History', *Victorian Studies*, XV, 1; Allan Nevins, *Ford*, 1954–62; George Ewart Evans, *Where Beards Wag All*, the foundry, migrant brewery labour, craftsmen. For fishing: George Ewart Evans, *The Days That We Have Seen*, 1975; Trevor Lummis, *Occupation and Society: the East Anglian Fishermen 1880–1914*, Cambridge, 1985; Paul Thompson with Tony Wailey and Trevor Lummis, *Living the Fishing*, 1983.

On pre-colonial African economies: Robert Harms, *River of Wealth, River of Sorrow: the Central Zaïre Basin in the era of the slave and ivory trade, 1500–1891*, New Haven, 1981; Steven Feierman, *The Shambaa Kingdom*, Madison, 1974, ch. 5; Roy Willis, *A State in the Making*, Bloomington, 1981; William Beinart, *The Political Economy of Pondoland*, Cambridge, 1982, ch. 5.

On agriculture: George Ewart Evans, *The Horse in the Furrow*, 1960, *The Farm and the Village*, 1969 and *Where Beards Wag All*; David Jenkins, *The Agricultural Community in South West Wales at the turn of the 20th Century*, 1971; Carol Faiers, 'Persistence and Change in Farming Methods in a Suffolk Village', *OH*, 4, 2. See also rural social history below.

On the entrepreneur: Thea Vigne, SSRC Research Report, 'Middle and Upper Class Families in the early 20th Century'; Jan and Ray Pahl, *Managers and their Wives*, 1971; single examples are Nevins, *Ford*, and George Ewart Evans, *From Mouths of Men*, pp. 21–30—on a Yorkshire wool manufacturer—and especially Carl B. Klockers, *The Professional Fence*, 1975.

On small businessmen: T. Vigne and A. Howkins, 'The small shopkeeper in industrial and market towns', in G. Crossick (ed.), *The Lower Middle Class in Britain 1870–1914*, 1977; Daniel Bertaux, 'The bakers of Paris', in *Our Common History*, and with Isabelle Bertaux-Wiame, 'Artisan Bakery in France: How It Lives and Why It Survives', in F. Bechhofer and B. Elliott (eds.), *The Petite Bourgeoisie*, 1981; Robert Scase and Robert Goffee, *The Real World of the Small Business Owner*, 1980; John Benson, *The Penny Capitalists: a Study of Nineteenth-Century Working-Class Entrepreneurs*, 1983.

History of science: David Edge, *Astronomy Transformed: The Emergence of Radio Astronomy in Britain*, 1977; Saul Benison, *Tom Rivers; Reflections on a Life in Medicine and Science*, Cambridge, Massachusetts, 1967; and Chris Niblett, 'Oral testimony and the social history of technology', *OH*, 8, 2.

History of religion: Robert Moore, *Pit-men, Preachers and Politics*, 1974; Hugh McLeod, 'White Collar Values and the Role of Religion' in *The Lower Middle Class in Britain* (see above), and 'Religion: the oral evidence', *OH*, 14, 1; Robert Towler and Audrey Chamberlain, 'Common Religion', *Sociological Year Book of Religion in Britain*, 6, 1973. Also the immense secondary literature in folklore—see references in Chapter 2 and below.

Labour history: for a full bibliography see Robert Turner, 'The contribution of oral evidence to labour history', *OH*, 4, 1; for annual updates, *BSSLH*.

Labour biography: John Saville and Joyce Bellamy, *Dictionary of Labour Biography*, from 1972; Alice and Staughton Lynd, *Rank and File: Personal Histories by Working Class Organisers from America*, New York, 1973; Paul Buhle, 'Radicalism, the oral history contribution', *IJOH*, 2, 3; People's Autobiography of Hackney, *Working Lives, 1905-45*, 1975; Arthur Randell, *Fenland Railwayman*, 1968; Angus Maclellan, *The Furrow Behind Me*, 1962; Margaret Powell, *Below Stairs*, 1968; and Angela Hewins, *The Dillen*, 1981.

On unionization, strikes, and campaigns: Peter Friedlander, *The Emergence of a UAW Local 1936-1939*, Pittsburgh, 1975; Anthony Mason, *The General Strike in the North-East*, 1970, and Peter Wyncoll and Hywel Francis in Jeffrey Skelley (ed.), *The General Strike*, 1976; Alun Howkins, 'Structural conflict and the farmworker: Norfolk 1900-20', *Journal of Peasant Studies*, 4, 3, 1977, and *Poor Labouring Men*, 1985; Joanna Bornat, 'Home and Work', *Women's History Issue, OH*, 5, 2; Steve Tolliday, 'Women and motor trades unions' and Fred Lindop, 'Dockers unofficial militancy, 1945-67', *Labour History Issue, OH*, 11, 2; Hywel Francis, *Miners against Fascism: Wales and the Spanish Civil War*, 1984; A. Lane and K. Roberts, *Strike at Pilkingtons*, 1971; R. Hay and J. McLaughlin, 'The Oral History of Upper Clyde Shipbuilders'; Sidney Pollard and Robert Turner, 'Profit-sharing and Autocracy: the case of J.T. and J. Taylor of Batley, Woollen Manufacturers, 1892-1966', *Business History*, 8, 1, January 1976.

On the unemployed: Studs Terkel, *Hard Times*, 1970; Barry Broadfoot, *Ten Lost Years*, Toronto, 1973; Dennis Marsden and E. Duff, *Workless*, 1973; Ray Broomhill, 'Using Oral Sources in writing a social history of the unemployed during the Depression in South Australia', in Joan Campbell (ed.), *Oral History 75*, Latrobe, 1975; Wendy Lowenstein, *Weevils in the Flour*, Melbourne, 1978.

On labouring communities: Raphael Samuel, ' "Quarry Roughs": Life and Labour in Headington Quarry, 1860–1920', *Village Life and Labour*, 1975; Moore, *Pit-men, Preachers and Politics*, and Robert Waller, *The Dukeries transformed: the social and political development of a twentieth-century coalfield*, Oxford, 1983; Jeremy Brecher, *Brass Valley: the story of working people's lives and struggles in an American industrial region*, Philadelphia, 1982. On fishing communities: Paul Thompson, *Living the Fishing*, 1983, and Trevor Lummis, *Occupation and Society*, Cambridge, 1985.

On work experience and class-consciousness: Studs Terkel, *Working*, New York, 1974; Luisa Passerini, *Torino operaia e fascismo*, Rome, 1984, and 'Work ideology and working-class attitudes to Fascism', in *Our Common History*; Alessandro Portelli, *Biografia di una Città*, Turin, 1985, and 'The time of my life', *IJOH*, 2, 3; Tamara Hareven, *Amoskeag*, New York, 1978 and *Family Time and Industrial Time*, Cambridge, 1982; Pierrette Pézerat and Danièle Poublan, 'French telephone operators', *OH*, 13, 1, and Cristina Borderias, 'Idendidad femenina y cambio social en Barcelona entre 1920 y 1980', in Mercedes Vilanova *et al.*, *El Poder en la Sociedad*, Barcelona, 1986. On fishermen, Thompson and Lummis above. Paul Thompson, 'Playing at skilled men: factory culture and pride in work skills among Coventry car workers', *Social History*, 1988; Frank McKenna, 'Victoria Railway Workers', *HW*, 1; Bob Gilding, *The Journeymen Coopers of East London*, 1971; Mavis Waters, 'Craft Consciousness in a Government Enterprise: Medway Dockyardmen, 1860–1906', Geoffrey Tyack, 'Service on the Cliveden Estate between the Wars', *OH*, 5, 1; Ronald Fraser, *In Search of a Past: the Manor House, Amnersfield, 1933–1945*, 1984; Dorothee Wierling, 'Women domestic servants in Germany at the turn of the century', *OH*, 10, 2. On German miners, Niethammer below; on British miners, David Douglass, *Pit Life in County Durham*, 1971 and 'The Durham Pitman' in R. Samuel (ed.), *Miners, Quarrymen and Saltworkers*, 1977; and George Ewart Evans, *From Mouths of Men*.

On the politics of the unorganized and rank and file in Britain: Jeremy Seabrook, *What went wrong? Working people and the ideals of the labour movement*, 1978, and Paul Thompson, *The Edwardians*, 1975, pp. 237 ff.

Political history: on the Chinese revolutions, William Hinton,

*Fanshen: a documentary of revolution in a Chinese village*, 1966, and *Shenfun: the continuing revolution in a Chinese village*, 1983; on the Mexican revolution, Oscar Lewis, *Pedro Martinez*, 1964, and the work of Mexico's national oral history programme led by Eugenia Meyer.

On European Fascism and Communism from the 1930s to the 1950s: Ronald Fraser, *Blood of Spain: the experience of Civil War, 1936-9*, 1979, and *In Hiding: the life of Manuel Cortes*, 1972; Jerome Mintz, *The Anarchists of Casa Viejas*, Chicago, 1982; Hywel Francis, 'Welsh Miners and the Spanish Civil War', *Journal of Contemporary History*, 53, 3, 1970; Lena Inowlocki, 'Denying the past: right wing extremist youth in West Germany', *L S*, 1985; Cristina Borderias and Mercedes Vilanova, 'Catalan miners and fishermen, 1931-9', Marjan Swegman, 'Women in resistance organisations in the Netherlands', papers by Anna Bravo and David Ellwood, and Lutz Niethammer, in *Our Common History*; Luisa Passerini, above, Lutz Niethammer *et al.*, trilogy on the Ruhr, *'Die Jahre weiss man nicht . . .', 'Hinterher merkt man . . .',* and *'Wir Kriegen jetzt andere Zeiten . . .'*, Bonn, 1983-5; Lothar Steinbach, *Ein Volk, Ein Reich, Ein Glaube?* Bonn, 1983; H. R. Kedward, *Resistance in Vichy France*, Oxford, 1978.

On the holocaust: Anna Bravo and Daniele Jalla, *La Vita Offesa*, Milan, 1986; Elmer Luchterhand, 'Knowing and not knowing: involvement in Nazi genocide', in *Our Common History*; Anna Bravo, 'Italian women in the Nazi camps—aspects of identity in their accounts', *O H*, 13, 1, and Shawl Esh, *Yad Washem Studies on the European Jewry Catastrophe and Resistance*, Jerusalem, 1960; Anna Maria Bruzzone and L. B. Rolfi, *Le donne di Ravensbruck*, Turin, 1978; Claudine Vegh, *I Didn't Say Goodbye*, 1984. For Hiroshima survivors: R. J. Lipton, *Death in Life*, New York, 1968.

Political documentary: William Manchester, *The Death of a President*, 1967; W. H. Van Voris, *Violence in Ulster: an Oral Documentary*, Amherst, 1975. Political biography: Thomas Harry Williams, *Huey Long*, 1969; Dean Albertson, *Roosevelt's Farmer: Claude R. Wickard in the New Deal*, New York, 1961; Merle Miller, *Plain Speaking: an oral biography of Harry S. Truman*, 1974; Valentina da Rocha Lima, *Getulio: una historia oral*, Rio de Janeiro, 1986; John Toland, *Adolf Hitler*, New York, 1976; Bernard Donoughue and George Jones, *Herbert Morrison: Portrait of a Politician*, 1973.

On military history: Mary P. Motley, *The Invisible Soldier: the experience of the Black soldier, World War II*, Detroit, 1975; General James Collins, Oscar Fitzgerald, Colonel Robert Zimmerman, and Benis Frank, 'Taped Interview and the Documentation of Viet Nam

Combat Operations', *OHR*, 1974; Lyn Macdonald, *Somme*, 1983; Alexander McKee, *Dresden 1945, The Devil's Tinderbox*, 1982. There are special research collections at the Imperial War Museum and National Maritime Museum, London, and Sunderland Polytechnic.

Colonial history: Charles Allen, *Plain Tales from the Raj*, 1975, and *Tales from the Dark Continent*, 1979; *Cambridge South Asian Archive*, 1973; J. J. Tawney, 'The Oral History programme of the Oxford Colonial Records Project', *OH*, 1, 3.

Life stories of the colonized: James Freeman, *Untouchable*, 1979; A. M. Kiki, *Kiki: 10,000 years in a lifetime*, 1968; Mary Smith, *Baba of Karo*, 1954; Edward Winter, *Beyond the Mountains of the Moon: the lives of four Africans*, 1959.

African history: for brief bibliographical reviews, see A. D. Roberts, 'The use of oral sources for African history', and Terence Ranger, 'Personal reminiscence and the experience of the people in East Central Africa', *OH*, 4, 1 and 6, 1; more extensively, Vansina and Henige, and the collected essays in Joseph Miller (ed.), *The African Past Speaks*, Folkestone, 1980. Bethwell Ogot, *History of the Southern Luo*, Nairobi, 1967, is a pioneering early study of pre-colonial history. Subsequently, David W. Cohen, *Womunafu's Bonafu*, Princeton, 1970 and *The Historical Tradition of the Busoga*, Oxford, 1972; Andrew Roberts, *A History of Bemba*, Madison, 1973; Steven Feierman, *The Shambaa Kingdom*, Madison, 1974; John Lamphear, *The Traditional History of the Jie of Uganda*, Oxford, 1976; Joseph Miller, 'The dynamics of oral tradition in Africa', in Bernardi, *Fonti orali*; Roy Willis, *A State in the Making: Myth, History and Social Transformation in Pre-colonial Ufipa*, Bloomington, 1981; and Paul Irwin, *Liptako Speaks: History from Oral Tradition in Africa*, Princeton, 1981.

On more recent social history and political struggle, see Don Barnett and Karari Njama, *Mau Mau from Within*, 1966; Terence Ranger, *Peasant Consciousness and Guerrilla Resistance in Zimbabwe*, 1985; William Beinart, *The Political Economy of Pondoland*, Cambridge, 1982; and Belinda Bozzoli (ed.), *Town and Countryside in the Transvaal*, Johannesburg, 1983.

Social history: general, Paul Thompson, *The Edwardians: the Remaking of British Society*, 1975.

Rural social history and community studies—the British south and east: George Ewart Evans, *Ask the Fellows who Cut the Hay*, 1956, *Where Beards Wag All, The Days That We Have Seen*, 1975; Ronald Blythe, *Akenfield: Portrait of an English Village*, 1969; Raphael Samuel, 'Quarry Roughs' in *Village Life and Labour*, above; Mary

Chamberlain, *Fenwomen*, 1975; Michael Winstanley, *Life in Kent at the turn of the century*, Folkestone, 1978. On the west, Scotland and Ireland: C.M. Arensberg and S.T. Kimball, *Family and Community in Ireland*, Cambridge, Massachusetts, 1940 and a critical re-study, Hugh Brody, *Inishkillane*, 1973; W.M. Williams, *The Sociology of an English Village: Gosforth*, 1956, and *A West Country Village: Ashworthy*, 1963; James Littlejohn, *Westrigg*, 1963; Ian Carter, *Farm life in north-east Scotland, 1840–1914*, 1979; Paul Thompson, *Living the Fishing*, 1983; Eric Cregeen, 'Oral Sources for the Social History of the Scottish Highlands and Islands' and 'Oral Tradition and Agrarian History in the West Highlands', *OH*, 2, 2, and 2, 1; David Jenkins, *The Agricultural Community in South West Wales*, above; I. Emmett, *A North Wales Village*, 1964; A.D. Rees, *Life in a Welsh Countryside*, Cardiff, 1950; R. Frankenberg, *Village on the Border*, 1957.

On the European peasantry: Fañch Elégoët on Brittany, Dagfinn Slettan on farmhands and farmwives in Norway, Anna Bravo on 'Italian peasant women in the First World War', and Gerhard Wilke on German village houses in *Our Common History*; Ronald Fraser, *El Pueblo, a mountain village on the Costa del Sol*, 1973; Anna Bravo, 'Solidarity and loneliness: Piedmontese peasant women at the turn of the century', *IJOH*, 3, 2; Nuto Revelli, *Il mondo dei vinti*, Turin, 1977 and *L'anello duro*, Turin, 1985; Alain Prévost, *Grenadou: paysan français*, Paris, 1966; Serge Grafteaux, *Mémé Santerre*, Paris, 1975; and Pierre Jakez Hélias, *Le cheval d'orgueil*, Paris, 1975 (*The Horse of Pride*, New Haven, 1978).

On peasants in the Third World and the Americas: see Africa above, and Aspasia Camargo on Latin America, *LS*, 1985; American anthropological life stories, ch. 2, note 36; Black history, below.

Urban social history and community studies—the city and city neighbourhood: Studs Terkel, *Division Street: America*, New York, 1967; Jerry White, below; partly autobiography, Robert Roberts, *The Classic Slum*, Manchester, 1971; Paul Thompson, 'Voices from Within', in Michael Wolff and H.J. Dyos (eds.), *The Victorian City*, 1973; People's Autobiography of Hackney series (*Working Lives*, 1975, *The Island*, 1979, etc.) from Centreprise, 136 Kingland High St., London E8; David Russell and George Walker, *Trafford Park*, 1979, from Manchester Studies, Manchester Polytechnic.

The smaller town: Melvyn Bragg, *Speak for England*, 1976; Raphael Samuel, 'Oral History and Local History', *History Workshop*, 1; Margaret Stacey, *Tradition and Change: A Study of Banbury*, 1960; Angela Hewins, *The Dillen*, 1981; Jeremy Seabrook, *City Close-up*,

1971; Robert and Helen Lynd, *Middletown, 1929, Middletown in Transition, 1937*, and Theodore Caplow, Reuben Hill, *et al.*, *Middletown Families: fifty years of change and continuity*, Minneapolis, 1982; Tamara Hareven and Randolph Langenbach, *Amoskeag: Life and Work in an American Factory City in New England*, 1979. Many sociological urban community studies also make some use of oral history, e.g. C. Rosser and C.C. Harris, *The Family and Social Change: A Study of Family and Kinship in a South Wales Town*, 1965.

On architecture, space and community: Jerry White, *Rothschild Buildings: Life in an East End Tenement Block 1887–1920*, 1980; and *The Worst Street in North London*, 1986; Henry Glassie, *Passing the Time in Ballymenone: culture and history of an Ulster Community*, Philadelphia, 1983; George McDaniel, *Hearth and Home: preserving a people's culture*, Philadelphia, 1983; and 'City Space and Order', *OH* special issue, 14, 2; Jerry White, 'Campbell Bunk', *HW*, 8 (1979).

Cultural history—education and language: Richard Hoggart, *The Uses of Literacy*, 1957: Brian Jackson and Dennis Marsden, *Education and the Working Class*, 1962; John Dillard, *Black English*, New York, 1973; Stanley Ellis, 'The Survey of English Dialects and Social History', *OH*, 2, 2, pp. 37–43; *Language and Class Workshop: Lore and Language*, the journal of the Sheffield University Centre for English Cultural Tradition and Language; Bruce Rosenberg, *The Art of the Folk Preacher*, 1970; Iona and Peter Opie, *The Lore and Language of School Children*, 1959. Urban folklore: Roger Abrahams, *Deep Down in the Jungle*, Chicago, 1970; rural: George Ewart Evans and George Thomson, *The Leaping Hare*, 1972; Eliot Wigginton (ed.), *Foxfire One, Two*, etc., New York, 1972—.

Music and song: Brian Jackson and Dennis Marsden, *Working Class Community* 1971—brass bands; A.L. Lloyd, *Folk Song in England*, 1967; David Buchan, *The Ballad and the Folk*, 1972; Lawrence Levine, *Black Culture and Black Consciousness*, New York, 1977; Edward Ives, *Joe Scott*, Urbana, 1978; Roy Palmer on Arthur Lane and George Dunn, *OH*, 8, 2, and 11, 1–2; Robin Morton, on John Maguire of Fermanagh, *Come Day, Go Day, God Send Sunday*, 1973; Gareth Stedman-Jones, 'Working-Class Culture and Working Class Politics in London, 1870–1900', *Journal of Social History*, 7, 4, 1974; Alun Howkins, 'The Voice of the People: the Social Meaning and Context of Country Songs' and Elizabeth Bird, 'Jazz Bands of North East England: the evolution of a working-class cultural activity', *OH*, 3, 1, and 4, 2; William J. Schafer, *Brass Bands and New Orleans Jazz*, Baton Rouge, 1977; Vivian Perlis, *Charles Ives Remembered*, New Haven, 1974;

Daniele Jalla, *La Musica: storia di una banda e dei suoi musicanti: Piossasco 1848–1980*, Turin, 1980.

Other leisure: Sally Alexander, *St Giles Fair 1830–1914*, History Workshop pamphlet 2, 1970; Michael Winstanley, 'The Rural Publican and his Business in East Kent before 1914', Harry Goldman, 'Worker's Theatre to Broadway Hit', and Bob Dickinson, 'Lancashire Music Halls' (*OH*, 4, 2; 10, 1; and 11, 1); Mike Steen, *Hollywood Speaks*, New York, 1974; Lawrence S. Ritter, *The Glory of their Times: the Story of the Early Days of Baseball*, New York, 1966; Gerald Sider, 'The ties that bind', *Social History*, 5, 1980.

Family history: Tamara Hareven, *Family Time and Industrial Time*, 1982; Elizabeth Roberts, *A Woman's Place: an oral history of working-class women, 1890–1940*, Oxford, 1984; Thompson, *Edwardians*; Daniele Jalla, 'The working-class family in Turin', in *Our Common History*; family issue, *Fonti orali*, autumn 1982; E. Le Roy Ladurie, *Montaillou*, Paris, 1975 (trans. 1978).

On childhood: Thea Thompson, *Edwardian Childhoods*, 1981 and (ed.) *OH*, 3, 2, *Family History Issue*: Thea Vigne, 'Parents and Children, 1890–1918: Distance and Dependence' and Elizabeth Roberts, 'Learning and Living: Socialisation outside School'; Jeremy Seabrook, *Working-class Childhood*, 1982. On youth: Stephen Humphries, *Hooligans or Rebels?*, Oxford, 1981; Paul Thompson, 'The War with Adults', Derek Thompson, 'Courtship and Marriage in Preston between the Wars' and Stephen Caunce, 'East Riding Hiring Fairs', *Family History Issue*. On marriage, sexuality, and birth control: Diana Gittins, 'Married Life and Birth Control between the Wars', *Family History Issue* and *OH*, 5, 2, and *Fair Sex: family size and structure 1900–39*, 1982; John R. Gillis, *For Better, For Worse: British marriages 1600 to the present*, Oxford, 1985; T. C. Smout, 'Aspects of Sexual Behaviour in 19th Century Scotland', in Maclaren, *Social Class in Scotland*; K. H. Connell, *Irish Peasant Society*; Martine Segalen, *Love and Power in the Peasant Family*, Oxford, 1983.

From anthropology: Oscar Lewis, *The Children of Sanchez*, New York, 1961. From sociology, community studies above, and also P. Townsend, *The Family Life of Old People*, 1957, and Lee Rainwater, *And the Poor Get Children*, 1960, and *Family Design*, Chicago, 1965; Mirra Komarovsky, *Blue Collar Marriage*, New York, 1962; also Jeremy Seabrook, *The Unprivileged*, 1967.

Women's history: Chamberlain, *Fenwomen*; Roberts, *A Woman's Place: an oral history of working-class women*; Sheila Rowbotham and Jean McCrindle, *Dutiful Daughters*, 1977; Jill Liddington and Jill

Norris, *One Hand Tied Behind Us*, 1978 (suffragists); *Women's History Issue, OH*, 5, 2, including articles on working-class women in the home and at work in Birmingham, Nottingham, Yorkshire, and Lancashire, by Catherine Hall, Sandra Taylor, Joanna Bornat, Elizabeth Roberts, and Jill Liddington; Second *Women's History Issue, OH*, 10, 2, including Angela John on women miners, Rickie Burman on Jewish wives, and Dorothee Wierling on German servants; Leonore Davidoff and Belinda Westover (eds.), *Our Work, Our Lives, Our Words: Women's History and Women's Work*, 1986; Amrit Wilson, *Finding a Voice*, below; Paul Thompson, 'Women in the Fishing: the roots of power between the sexes', *Comparative Studies in Society and History*, 27 (1985), 1; Sherna Gluck, 'Interlude or change: women and the World War II experience, *IJOH*, 3, 2, and *From Parlor to Prison: Five American Suffragettes Talk about Their Lives*, New York, 1976; Gwendolyn Safier, *Contemporary American Leaders in Nursing: an oral history*, New York, 1977; Regina Markell Morantz *et al., In Her Own Words: oral histories of women physicians*, New Haven, 1982; 'Women's history and oral history', *Frontiers*, II, 2 (1977); Jan Carter, *Nothing to Spare: Recollections of Australian Pioneering Women*, Melbourne, 1981; Anna Bravo, 'Solidarity and loneliness; Piedmontese peasant women', *IJOH*, 3, 2, and 'Italian women in the Nazi camps', *OH*, 13, 1; Bravo, 'Italian peasant women and the First World War', Nelleke Bakker on Amsterdam seamstresses, etc., in *Our Common History*; Nuto Revelli, *L'Anello duro*, 1985; *Fonti orali*, women's history issue, autumn 1981.

On criminal and deviant subcultures: James Bennett, *Oral History and Delinquency*, Chicago, 1982 and other Chicago works, ch. 2 above; Helen Hughes, *The Fantastic Lodge: the autobiography of a girl drug addict*, 1961; R. Bogdan, *Being Different*, 1974; Tony Parker and Robert Allerton, *The Courage of His Convictions*, 1962, Tony Parker, *The Unknown Citizen*, 1963, and *The Twisting Lane*, 1969; Walter Probyn, *Angel Face: the making of a criminal*, 1977; Raphael Samuel, *East End Underworld: Chapters in the Life of Arthur Harding*, 1981; Steve Humphries, *Hooligans or Rebels?*; Jerry White, 'Police and people in London', *OH*, 11, 2.

Minorities—American Indians: Paul Radin, *Crashing Thunder*, 1926; Leo Simmons, *Sun Chief*, New Haven, 1942; Nancy O. Lurie, *Mountain Wolf Woman: the Autobiography of a Winnebago Indian*, Ann Arbor, 1961; Frank Waters, *Book of the Hopi*, 1963; J.H. Cash and H.T. Hoover (eds.), *To Be an Indian*, New York, 1971; Howard and Frances Morphy, 'The "myths" of Ngalakan history: ideology and images of the past in Northern Australia', *Man*, 19 (1984), 1.

Chinese and Japanese: Victor G. and Brett de Barry Nee, *Longtime Californ': a documentary Study of an American Chinatown*, New York, 1973; Diana Marlatt, *Steveston Recollected*, Victoria B.C., 1975.

Jewish history: Sydelle Kramer and Jenny Masur, *Jewish Grandmothers*, Boston, 1976; William J. Fishman, *East End Jewish Radicals, 1875–1914*, 1975; Bill Williams, 'The Jewish Immigrant in Manchester', and Rickie Burman, 'The Jewish woman as breadwinner', *OH*, 7, 1 and 10, 2; Jerry White, *Rothschild Buildings*, 1980.

On migration: Isabelle Bertaux-Wiame, 'The life history approach to the study of internal migration', in *Our Common History*; Edward Orser, 'Racial change in retrospect', *IJOH*, 5, 1; Jane Synge, 'Immigrant communities in early twentieth-century Hamilton, Canada', Gina Harkell, 'The migration of mining families to the Kent coalfield', and Margaret Mackay, 'Tiree emigrant communities in Ontario, Canada', *OH*, 4, 2; 6, 1; and 9, 2; Wendy Lowenstein, *The Immigrants*, Melbourne, 1978; Joan Morrison and Charlotte Zabusky, *American Mosaic*, New York, 1980.

Black history—urban: Paul Bullock, *Watts, the Aftermath*, New York, 1969; Alex Haley, *Autobiography of Malcolm X*, New York, 1965. Rural: *These Are Our Lives*, Federal Writers Project, p. 59 above; Theodore Rosengarten, *All God's Dangers: The Life of Nate Shaw*, New York, 1974; William Montell, *The Saga of Coe Ridge*, Knoxville, Tennessee, 1970; Lawrence Goodwin, 'Populist Dreams and Negro Rights: East Texas as a Case Study', *American Historical Review*, 76, 1, 1971; Scott Ellsworth, *Death in a Promised Land: The Tulsa Race Riot of 1921*, Baton Rouge, 1982; *Southern Exposure*, special issue 'Voices of Southern Struggle', 1, 3–4, Winter 1974; George Rawick, *From Sundown to Sunup, The Making of the Black Community* and *The American Slave—A Composite Biography*, Westport, Connecticut, 1972. Earlier extracts from the slave narratives were published in B.A. Botkin, *Lay My Burden Down*, Chicago, 1945; for a recent use, Eugene Genovese, *Roll, Jordon Roll: the World the Slaves Made*, 1974. From sociology: especially Charles S. Johnson, *Shadow of the Plantation*, Chicago, 1934; John Dollard, *Caste and Class in a Southern Town*, New Haven, 1937; John K. Morland, *Millways of Kent*, Chapel Hill, 1958, and Hylan Lewis, *Blackways of Kent*, Chapel Hill, 1955. For Britain: Amrit Wilson, *Finding a Voice: Asian Women in Britain*, 1978; and *Black History Issue, OH*, 8, 1—Harry Goulbourne, 'Oral history and black labour', Elizabeth Thomas-Hope, 'West Indian Migration to Britain', Pnina Werbner, 'A Community of Suffering', and Donald Hinds on Brixton.

1. George Ewart Evans, *The Days That We Have Seen*, p. 24.
2. 'The Miners, the relevance of oral evidence', *OH*, 1, 4, p. 74.
3. pp. xxxii–xxxviii.
4. Donoughue, *OH*, 2, 1, pp. 83–4.
5. p. 4.
6. pp. 127, 144, 201.

## Chapter 4 Evidence

On memory: especially F. C. Bartlett, *Remembering*, Cambridge, 1932; Ian Hunter, *Memory*, 1957, revised edition 1964; Roberta L. Klatsky, *Human Memory*, San Francisco, 1975.

On historical changes in perception: Mandrou, *Introduction to Modern France*; also, generally, Vansina, above (both original and revised editions). Gerontology: A. T. Welford, *Ageing and Human Skill*, 1958; B. L. Neugarten (ed.), *Middle Age and Ageing*, 1968. On interviewing and sociological methods: Herbert H. Hyman, *Interviewing in Social Research*, Chicago, 1954; W. V. D. Bingham and B. V. Moore, *How to Interview*, 1931, revised New York, 1959; Norman Denzin (ed.), *Sociological Methods*, Berkeley, 1970; Barney G. Glaser and Anselm L. Strauss, *The Discovery of Grounded Theory*, Chicago, 1967. On consistency of accuracy in memory, in addition to works cited in notes, Alan Baddeley, Diana Gittins, and Colin Hindley in Louis Moss and Harvey Goldstein (eds.), *The Recall Method in Social Surveys*, 1979.

On the influence of inner consciousness and identity—'subjectivity'—and mutual interaction within the interview in shaping memory, concepts, form, and language, see especially Ronald Fraser, *In Search of a Past*, 1984; Elliot Mishler, *Research Interviewing: Context and Narrative*, Cambridge, Mass., 1986; Luisa Passerini, *Torino operaia e fascismo*, 1984; Luisa Passerini and Isabelle Bertaux-Wiame in *Our Common History*; Anna Bravo, 'Italian women in the Nazi camps', *OH*, 13, 1; Alessandro Portelli, 'The peculiarities of oral history', *HW*, 12, and 'Functions of time in oral history', and reply to Louise Tilly, *IJOH*, 2, 3, and 6, 1.

1. *OH*, 1, 4, p. 93.
2. pp. 133–7.
3. D. Read, *Documents from Edwardian England*, 1973, pp. 305–7; Robert Blake, *The Unknown Prime Minister, the Life and Times of Bonar Law*, 1955, p. 130.
4. 'Populist Dreams', *American Historical Review 76*, above; cf. Mason, *General Strike . . .*, pp. 70, 101.
5. *OH*, 1, 3, pp. 35, 46.

6. Trans. 1952, p. 37; Jack Douglas, *The Social Meanings of Suicide*, Princeton, 1967.
7. Census of England and Wales for 1911, Volume XIII, 'Fertility and Marriage', p. xv.
8. Frances Widdowson, 'Elementary Teacher Training and the Middle Class Girl', Susan Miller, 'The Happy Coincidence: rural poverty and the labour migration scheme of 1835-7', and Eve Hostettler, 'Cottage Economy', University of Essex M.A. dissertations and Oral History projects, 1976; Gudie Lawaetz, 'History on Film', *History Workshop*, 2, p. 124, cf. p. 137; Royden Harrison, *Times Higher Education Supplement*, 23 July 1976.
9. R.H.S. Crossman, *Listener*, 1 February 1973, p. 148; *From Mouths of Men*, pp. 174-5.
10. See Alessandro Portelli, 'Oral history, the law and the making of history', *HW*, 20, pp. 5-25, on the Aldo Moro trial of the Italian Red Brigades.
11. Hunter, p. 175.
12. Re-recording, e.g. Calum Johnston, recorded by Alan Lomax of the School of Scottish Studies in 1951, *Tocher*, 13, p. 166, and for our own survey, 293, pp. 1-8, 1971; Thea Vigne, analysis of three re-interviews for Open University course D 301, Notes for Television Programmes (8 and 9), pp. 19-23; and *OH*, 1, 4, pp. 14-17.
13. H.P. Bahrick, P.O. Bahrick, and R.P. Wittlinger, 'Fifty Years of Memory for Names and Faces', *Journal of Experimental Psychology* (104, 1), March 1975.
14. Jenkins, *The Agricultural Community in South West Wales*, p. 5; Hunter, p. 161.
15. Bartlett, p. 204.
16. 1, pp. 446-7.
17. Hunter, p. 227; Welford, p. 233.
18. Plummer, p. 103.
19. Mishler, pp. 18-19; Hyman, p. 115.
20. Hyman, pp. 159-61.
21. Vansina, p. 93.
22. Denzin, pp. 199-203.
23. Carol Faiers, 'Persistence and change in a Suffolk village', Essex University Sociology BA project, 1976, p. 36.
24. *OH*, 4, 1, p. 47.
25. 'Implications of Oracy: an Anthropological View', *OH*, 3, 1, pp. 41-9.
26. Denzin, p. 324; Hyman, pp. 234, 238-42.

27. pp. 459, 467.
28. Butler and Stokes, p. 273; Charles More, *Skill and the English Working Class 1870–1914*, 1980, pp. 58–61, 66–7, 70–1.
29. pp. 190–2.
30. Neugarten, pp. 173–7.
31. Herbert Blumer, *Critiques of Research in the Social Sciences: I: An Appraisal of Thomas and Znaniecki's The Polish Peasant in Europe and America* (New York, 1939), introduction to revised edition, New Brunswick, 1979, p. xxxiv; Yves Lequin and J. Metral, 'Une mémoire collective: les métallurgistes retraités de Givors', *Annales*, 35, 1.
32. Unpublished interview, Paul Thompson and Thea Vigne, 908, pp. 10–16; cf. *OH*, 1, 4, p. 31.
33. Mason and Skelley, on the General Strike; on Yad Washem, *OH*, 5, 1, pp. 37–9.
34. Raphael Samuel, *HW*, 1, p. 202 and *Miners, Quarrymen and Saltworkers*, 1977, p. 4.
35. Friedlander, *Emergence of a UAW Local*, p. xxx.
36. *OH*, 1, 4, pp. 116–20; Munson in *Contemporary Review*, 1975–6, pp. 107–8.
37. *OH*, 1, 3, p. 39; *OH*, 2, 1, p. 84.
38. *Speak for England*, p. 7.
39. p. 5.
40. *Ford*, 1, pp. 267–8, 389–93.
41. *Biografia di una Città: storia e racconto: Terni 1830–1985*, Turin, 1985, pp. 261, 307–8; *IJOH*, 2, 3, pp. 172, 175; and *HW*, 12, p. 100; cf. Vansina (1985), p. 6.
42. Roy Hay, 'Use and Abuse of Oral Evidence', unpublished paper.
43. *HW*, 12, p. 100.
44. John Berger, *Pig Earth*, 1985, p. 9.
45. Vansina (1985), pp. 21, 68–9.
46. Pierre Gaudin and Claire Reverchon, 'Le sens du tragique dans la mémoire historique', papers of the Aix 1982 International Oral History Conference, pp. 89–98; Françoise Zonabend, *La Mémoire Longue*, Paris, 1980, pp. 107–11, 299–304.
47. *OH*, 5, 1, p. 23.
48. Vansina (1985), pp. 56, 103; Paul Irwin, *Liptako Speaks*, 1981; Henige, pp. 72–3; Howard and Frances Morphy, 'The "myths" of Ngalakan history', *Man*, 19, 3 (1984), pp. 459–78.
49. Steven Feierman, *The Shambaa Kingdom*, 1974, p. 15; Anna Bravo and Daniele Jalla, '*La Vita Offesa*, Milan, 1986, p. 63; Passerini in

*Our Common History*, p. 61; Joutard, *Ces voix*, p. 235; Lucien Aschieri, *Le passé recomposé: Mémoire d'une communauté Provençale*, Marseilles, 1985.

50. Carolyn Steedman, *Landscape for a Good Woman*, 1986, p. 39; Vansina, *OH*, 5, 1, pp. 22–3; Ronald J. Grele, *Envelopes of Sound*, Chicago, 1975, pp. 57, 61.

51. F. A. Salome, 'The methodological significance of the lying informant', *Anthropological Quarterly*, 50 (1977), pp. 117–24; Vansina (1985), p. 8; Frank Coffield, P. Robinson, and J. Sarsby, *A Cycle of Deprivation? A case study of four families*, 1980, pp. 13–14, 33.

52. Jack Goody, *Literacy in Traditional Societies*, Cambridge, 1968, pp. 27–68; Vansina (1985), pp. xii, 92, 94, 120–3, 162–5.

53. Jerome Mintz, *The Anarchists of Casa Viejas*, Chicago, 1982, pp. xi, 271; Eric Hobsbawm, *Primitive Rebels*, 1959, p. 90.

54. James W. Wilkie, 'Alternative views in History: Historical statistics and oral history', in Richard E. Greenleaf and Michael C. Meyer (eds.), *Research in Mexican History*, Lincoln, Nebraska, 1973, p. 54; Vansina in Joseph Miller (ed.), *The African Past Speaks*, Folkestone, 1980, p. 276.

## Chapter 5 Memory and the Self

On psychoanalysis and memory, see the suggestions on 'subjectivity' for h. 4; Freud's own essays on 'The Interpretation of Dreams' (1900), Jokes and their Relation to the Unconscious' (1905), and 'The Ego and he Id' (1923), *Complete Psychological Works*, 1953—IV–V, VIII, and XIX; Juliet Mitchell, *Psychoanalysis and Feminism*, 1974; Nancy Chodorow, *The Reproduction of Mothering*, Berkeley, 1978; Gill Gorell arnes, 'Systems theory and family therapy', in M. Rutter and L. Lerzov (eds.), *Modern Child Psychiatry*, 1985, pp. 216–29; Jeremy Holmes, 'Family and individual therapy: comparisons and contrasts', *British Journal of Psychiatry*, 47 (1985), 668–76.

On reminiscence therapy, in addition to the articles below, *Recall—A Handbook*, Help the Aged, 1986; Jane Lawrence and Jane Mace, *Remembering in Groups*, Exploring Living Memory, 1987; Andrew orris, *Reminiscence with Elderly People*, 1986; Peter Coleman, *Elderly People and the Reminiscence Process*, 1986; and M. Kaminisky d.), *The Uses of Reminiscence: New Ways of Working with Older dults*, New York, 1984—special double issue, *Journal of Gerontological Social Work*, 7, 1–2.

. pp. 85, 118.

. *Our Common History*, pp. 192–3.

3. Sigmund Freud, 'Remembering, repeating and working-through' (1914), *Complete Psychological Works*, 1966–74, 12, p. 148; David Lowenthal, *The Past is a Foreign Country*, 1985, p. 17.

4. Donald and Lorna Miller, 'Armenian Survivors: a typological analysis of victim response', *OHR*, 1982, pp. 47–72.

5. Anna Bravo and Daniele Jalla, *La Vita Offesa*, Milan, 1986, p. 160; Claudine Vegh, *I Didn't Say Goodbye*, 1984, pp. 29, 161.

6. Ponsonby, pp. 8–9; Baum, *International Journal on Aging and Human Development*, 12 (1980–1), pp. 49–53; Faraday and Plummer, 'Doing Life Histories', *Sociological Review*, 27 (1979), pp. 773–98.

7. Peter Coleman, 'Issues in the therapeutic use of reminiscence with elderly people', in Ian Hanley and Mary Gilhooly (eds.), *Psychological Therapies for the Elderly*, 1986, pp. 41–64; R. Dobrof, introduction to M. Kaminsky (ed.), *The Uses of Reminiscence: New Ways of Working with Older Adults*, New York, 1984; Robert Butler, 'The Life Review: an Interpretation of Reminiscence in the Aged', *Psychiatry*, 26 (1963), pp. 65–76, 'The Life Review: an Unrecognized Bonanza', *International Journal on Aging and Human Development*, 12 (1980–1), 35–8, and *Why Survive?*, New York, 1975, pp. 412–4.

8. S. Merriam, 'The concept and function of reminiscence: a review of the research', *Gerontologist*, 20 (1980), 604–9; V. Molinari and R.E. Reichlin, 'Life review reminiscence in the elderly: a review of the literature', *International Journal on Aging and Human Development*, 20 (1984–5), 81–92.

9. Debbie Frost and Kay Taylor, 'Life Story Books: This is my life', *Community Care*, 7 August 1986; S.M. Hale (a social worker in Peckham), *Case Conference*, 7, 6 (1960), 153–5; Malcolm Johnson, 'That was your life: a biographical approach to later life', in Vida Carver and Penny Liddiard (eds.), *An Ageing Population*, 1979, pp. 147–61 (152, 159); Mel Wright, 'Using the past to help the present', *Community Care*, 533, 11 October 1984, and 'Priming the past', *OH*, 14, 1.

10. John Adams, 'Reminiscence in the geriatric ward: an undervalued resource', *OH*, 12, 2; J. Lesser *et al.*, 'Reminiscence group therapy with psychotic geriatric patients', *Gerontologist*, 21 (1981), 291–6; Andrew Norris and Mohammed Abu El Eileh, 'Reminiscence groups: a therapy for both elderly patients and their staff', *Nursing Times*, 78 (1982), 1368–9, or *OH*, 11, 1; Annie Lai, Bob and Pippa Little, 'Chinatown Annie: the East End opium trade 1920–35', *OH*, 14, 1.

11. Peter Coleman, in Hanley and Gilhooly, and 'The past in the present: a study of elderly people's attitudes to reminiscence', *OH*, 14, 1.
12. H. Hazan, *The Limbo People*, 1980, p. 27–8. The true source of this poem, which has been cited in several places, is unclear, but it has a suitably mythical origin in a London old people's centre.

## Chapter 6 Projects

For practical advice on projects generally: Stephen Humphries, *The Handbook of Oral History*, 1984.

For educational projects: Sallie Purkis, *Oral History in Schools*, Oral History Society, 1980, *Thanks for the Memory*, 1987, and regular reports in *OH*; also, Alistair Ross, 'Children as historians' and Elyse Dodgson, 'From oral history to drama', *OH*, 12, 2, and Liz Cleaver, 'Oral history at Thurston Upper School', *OH*, 13, 1; Elyse Dodgson, *Motherland*, 1984. Source books using oral history include a series for eight-year-olds, Sallie Purkis, *At Home in 1900*, etc., 1981; a model teaching pack, *Hurrah for Life in the Factory*, is available from Manchester Studies, Manchester Polytechnic; and tape sets available include BBC School Radio's *People Talking* and *From Scotland's Past*.

For American educational work, see John Neuenschwander, *Oral History as a Teaching Approach*, Washington, 1976; Eliot Wigginton, *The Foxfire Book*, New York, 1972. In higher education: James Hoopes, *Oral History: an Introduction for Students*, Chapel Hill, 1979; and on Third World field-work, David Henige, *Oral Historiography*, 1982.

For community projects: Humphries; Jane Mace, *Working with Words: Literacy beyond School*, 1979, and *Write First Time*; Willa Baum, *Oral History for the Local Historical Society*, Berkeley, revised edition, 1971, and Barbara Allen and Lynwood Montell, *From Memory to History: Using Oral Sources in Local Historical Research*, Nashville, 1981; David Lance, *An Archive Approach to Oral History*, 1978, and Willa Baum, 'The Expanding Role of the Librarian in Oral History', in Dunaway and Baum; Richard Gray on Peckham People's History, Graham Smith on the Arbroath MSC History Project, Robert Perks on Bradford Heritage and Sian Jones on Southampton Museum in *Oral History and Community History*, *OH*, 12, 2; Sian Jones, April Whincop, Elizabeth Frostick, *et al.* in *Museums and Oral History*, *OH*, 4, 2; George McDaniel, *Hearth and Home: Preserving a People's Culture*, Philadelphia, 1983; Rickie Burman, 'Participating in the past?

Oral History and Community History in the work of Manchester Studies', *IJOH*, 5, 2; and John Kuo Wei Tchen, 'Towards Building a Democratic Community Culture: reflections on the New York Chinatown History Project', Fifth International Oral History Conference, Barcelona, 1985.

1. Reprinted from *Write First Time*, June 1981, by permission of the author.
2. *Foxfire One*, 1972, pp. 141, 221, 363, 375; *OHR*, 1973, pp. 30–6.
3. A selection of these projects is published in Leonore Davidoff and Belinda Westover (eds.), *Our Work, Our Lives, Our Words*, 1986.
4. *HW*, 1, pp. 198–9.
5. *HW*, 1, pp. 1–2.
6. Humphries, p. 95.
7. *My Apprenticeship*, p. 362.
8. *HW*, 1, pp. 117–20.

## Chapter 7 The Interview

Full discussions of the interview method can be found in Hyman, *Interviewing in Social Research*, and Bingham and Moore, *How to Interview*, above. Beatrice Webb, *My Apprenticeship*, 1926, pp. 361–3, is still worth reading. More immediately helpful advice can be found in Baum, *Oral History for the Local Historical Society*, above; George Ewart Evans, 'Approaches to Interviewing', Michael Winstanley, 'Some Practical Hints on Oral History Interviewing', *OH*, 1, 4, and 5, 1. On recording tradition-bearers, see Donald A. Macdonald, 'Collecting Oral Literature', in R.M. Dorson (ed.), *Folklore and Folklife*, Chicago, 1972, pp. 407–30.

On interviewing politicians and other leaders: Anthony Seldon and Joanna Pappworth, *By Word of Mouth: Elite Oral History*, 1983; Peter Oliver, 'Oral History: One Historian's View', *COHA Journal*, 1; and James Wilkie, 'Alternative Views in History: Historical Statistics and Oral History', in Richard E. Greenleaf and Michael C. Meyer (eds.), *Research in Mexican History*, 1973. I have also found helpful Roy Hay's unpublished paper, 'Use and Abuse of Oral Evidence'.

1. Hay, p. 15.
2. *OHR*, 1976, p. 30.
3. *OH*, 3, 1, p. 21.
4. *My Apprenticeship*, pp. 361–2.
5. E.g. Michael Winstanley, field-work notes, University of Kent; Charles Parker, *OH*, 1, 4, pp. 53–4. 'Tradition-bearer' is a common term among folklorists; 'narrator' with American programmes.

6. *OH*, 1, 4, pp. 62–3.
7. Janet Askham, 'Telling stories', *Sociological Review*, 30 (1982), pp. 555–73. The completely unstructured initial interview was used in the early work of Luisa Passerini and her colleagues in Turin.
8. *Documents of Life*, p. 103.
9. Hay, pp. 13–14.
10. Baum, p. 33.
11. Hence Beatrice Webb's view, *My Apprenticeship*, p. 362; or Thomas Reeve's experience with liberal intellectuals, *OHR*, 1976, p. 33.
12. *OH*, 1, 4, p. 56.
13. Saville, *OH*, 1, 4, p. 56; Edge, 4, 2, p. 10; *My Apprenticeship*, p. 363.
14. *Tape Recording of Local Dialect*, Standing Conference for Local History, 1971.
15. p. 362.
16. Ann Oakley, 'Interviewing women: a contradiction in terms', in Helen Roberts (ed.), *Doing Feminist Research*, 1981, pp. 30–61. For her own earlier study of housework she had used an elaborate formal interview schedule, but investigating the experience of childbirth through repeated interviews and sometimes presence at the birth itself required a more flexible, intimate, and mutual approach.
17. p. 362.
18. Oliver, p. 14.
19. *OH*, 1, 3, pp. 35–8.
20. Oliver, p. 14.
21. Vansina, pp. 198–200.

## Chapter 8  Storing and Sifting

For preservation, storage, and archiving, see David Lance, *An Archive Approach to Oral History*, 1978. Also, on the legal situation in Britain: David Lance, *OH*, 4, 1, pp. 96–7; in America: Baum, *Oral History for the Local Historical Society*, and Truman Eustis, *OHR*, 1976, pp. 6–18. On transcribing, see Raphael Samuel, 'Perils of Transcript', *OH*, 1, 2, on which I have drawn closely. On choosing a word processor for transcription and cataloguing, general advice is given by Roy Rosensweig, 'Automating your oral history program: a guide to data base management on a microcomputer', *IJOH*, 5, 3; and Frederick J. Stielow, *The Management of Oral History Sound Archives*, New York, 1986. See also ch. 9, n. 14.

1. Technically, for the tapes alone, a temperature of 5° to 10° would be still better, but because the tapes would require up to twelve hours reacclimatization before use at ordinary office temperatures, this is only worth considering for master tapes intended for very rare use.

2. Baum, pp. 41–4; the Oral History Association guideline is less precise, p. 46; she writes of the American legal situation generally concerning oral history, pp. 47–8, that libel is defined as the publication 'of false utterance, without just cause, and which tends to expose the other to public hatred, contempt, or ridicule. Various court cases have progressively reduced the possibility of a court finding any historical effort either slanderous or libelous. First, the dead cannot be libeled. Second, libelous defamation of prominent living persons must include actual malice plus irresponsible disregard for truth.

   'The researcher or interviewer may need to be a little more concerned if his questioning leads him into the purely private lives of prominent or not-prominent persons, although he still has a good defense if he can indicate this is truth published for good motives. Of course, there always exists the possibility of harassment in the lower courts through filing of suits for defamation. Such suits stand almost no possibility of ending in court award for damages but they could cost time and expense to the researcher in defending himself. But to all intents and purposes, slander or libel is a non-existent danger to an oral history project. It is the project's reputation for responsible work that needs guarding, not its legal liability.'

   The question of literary copyright in oral history recordings remains as uncertain as in Britain. This statement is elaborated in an article by Truman W. Eustis III on American copyright law (*OHR*, 1976, pp. 6–18), with examples from cases including the important decision of New York State Court of Appeals in Ernest Hemingway's Estate *v* Random House (1968) not to prevent publication by Hemingway's writer friend A. E. Hotchner of conversations which he had noted (not taped) with him. This decision, that 'Ernest Hemingway impliedly licensed his rights under common-law copyright when he knowingly permitted Hotchner to interview him', brings the United States, according to James W. Wilkie, into line with the 'common-sense position' which 'appears to hold true throughout Latin America'—and certainly in Mexican procedure—that 'intellectual authorship is held by the interviewer' (*Research in Mexican History*, p. 55).

3. p. 32.

4. p. 212. In the same spirit, the Aural History Institute of British

Columbia, *Manual*, p. 40, advises the indication of local accent through spellings like 'yeah', 'huh', 'must'a', 'gonna'.

5. With of course an explanatory letter which seeks to avoid some of the difficulties which follow. The Aural History Institute's example (ibid., p. 49) includes this paragraph:

Please read the transcript, remembering that it is a record of the spoken, rather than the written word. Change any incorrect dates, misspelled names or misinformation; correction of grammar is not recommended as it would distort the oral record. If you discover that you have omitted information concerning specific instances, add it in the margin or on additional sheets of paper. Similarly, if you wish to clarify statements that you have made, please add the information to the transcript.

## Chapter 9 Interpretation: the Making of History

For analysis, and the role of theory in field-work, see Jan Vansina, 'The Power of Systematic Doubt in Historical Enquiry', *History in Africa*, 1, 1974, pp. 109–27; Peter Friedlander, introduction to *The Emergence of a UAW Local 1936–1939*, Pittsburgh, 1975; Martin Bulmer (ed.), *Sociological Research Methods*, 1977; and Paul Thompson, 'Life Histories and the Analysis of Social Change', in Bertaux, *Biography and Society*.

For literary forms of analysis, see Elliot Mishler, *Research Interviewing: Context and Narrative*, Cambridge, Mass., 1986; Walter Ong, *Orality and Literacy*, 1982; Ron Grele, 'Listen to their Voices', *OH*, 7, 1 and (ed.) *Envelopes of Sound*, Chicago, 1975; Isabelle Bertaux-Wiame in *Our Common History*; William Labov, *Language in the Inner City: Studies in the Black English Vernacular*, Philadelphia, 1972 (Oxford, 1977), ch. 5 and 9; Robert Fothergill, *Private Chronicles: a Study of English Diaries*, 1974 (esp. ch. 5); David Vincent, *Bread, Knowledge and Freedom: a study of nineteenth-century working-class autobiography*, 1981; Philippe Lejeune, *Je est un autre: L'autobiographie, de la littérature aux médias*, Paris, 1980; E. Culpepper Clark, Michael Hyde, and Eva McMahan, 'Communication in the oral history interview', and Alessandro Portelli, 'Functions of time in oral history', *IJOH*, 1, 1 and 2, 3.

For quantitative analysis: Richard Jensen, 'Oral History, Quantification and the New Social History', *OHR*, 1981; Trevor Lummis, 'Structure and validity in oral evidence', *IJOH*, 2, 2, and *Listening to History*, 1987, pp. 94–106.

1. Copying requires two tape recorders and a connecting lead. The

master tape is placed on one recorder and a clean tape on the second recorder, and the lead is inserted into the appropriate sockets so that the second recorder is recording sound from the first. No microphone is required; this is a direct process, and the sound from the tape on the first machine will be reproduced through the speaker of the second at the same time. The sound from the second speaker is simply a convenience, so that you know what is being copied, and you can play it high or low as you wish without affecting the quality of the copy. The quality depends upon how you adjust the volume of the *first* speaker, and the *recording level* of the *second* tape recorder. When you have found exactly the place for copying on *both* tapes, you simultaneously start playback on the first machine and recording on the second, stopping both at the end of the extract. Alternatively, you can start the second machine just before and stop it just after the first machine, so that there is a brief gap between extracts. If you are making a series of extracts, you then remove the first master tape, place the next on the first recorder, find the place you want, and copy as before. With a little dexterity, it is quite a quick process.

2. *OH*, 5, 1, p. 22.
3. Grele, 'Listen to their Voices'.
4. Passerini, pp. 17, 22, 33.
5. Ong, pp. 24, 111; Bertaux-Wiame.
6. Mishler, p. 68; Janet Askham, 'Telling stories', *Sociological Review*, 30 (1982); Carolyn Steedman, *Landscape for a Good Woman*, 1986, pp. 58–9; cf. Passerini, p. 16.
7. Passerini, p. 43; Stefan Bohman, 'The People's Story: on the collection and analysis of autobiographical materials', paper to conference on Working-Class Culture, Nörrköping, September 1986.
8. Mishler, pp. vii–viii, 18–19, 53 ff.
9. Michael Holquist (ed.), *The Dialogic Imagination: Four Essays by M. M. Bakhtin*, 1981, p. xx.
10. SSRC Research Report; cf. his *Occupation and Society* Cambridge, 1985, and *IJOH*, 2, 2.
11. *COHA Journal*, 1, pp. 28–9.
12. p. xiv.
13. Course D 301, radio programme, 'The Small Household'; Jensen, *OHR*, 1981.
14. For the development of life story computer programs (pioneered by Jorge Balan and Elizabeth Jelin in the late 1960s) see Stephen Tagg, 'Life Story interviews and their interpretation', in Michael

Brenner *et al.* (eds.), *The Research Interview*, 1985, pp. 182–7.

15. Annabel Faraday and Ken Plummer, 'Doing Life Histories', *Sociological Review*, 27 (1979), pp. 773–98; cf. *Documents of Life*, pp. 119 ff.

16. *The Poverty of Theory*, 1978, pp. 229–42.

17. pp. ix–xxxiii.

18. Hareven and Langenbach, *Amoskeag*, New York, 1978, pp. 381–8, and Hareven, *Family Time and Industrial Time*, Cambridge, 1982.

19. 'Home and Work: a New Context for Trade Union History', *OH*, 5, 2.

20. 'Women in the Fishing: the Roots of Power between the Sexes', *Comparative Studies in Society and History*, 27 (1985), 3–32.

21. Ch. 5.

22. *All God's Dangers*, pp. 453–4.

# Model Questions

These questions are *not* a questionnaire but an outline interviewer's guide—in the spirit of Chapter 7. The interviewer's directions are printed in *italics*. Where there is a question mark, the form of the question is as suggested; elsewhere, points for questioning are in a summarized form and need expanded wording in use.

## 1. The Household: Basic Information

Informant's name, present address, year of birth, marital status, year of marriage, birthplace (street or district if known).

How many years did you live in the house where you were born? Where did you live then? *Continue for later moves.* Do you remember why the family made these moves? *If family moved a significant distance*: who helped at each end, the journey, first impressions, economic effect of move; continuing contact with original home and other migrants.

How many brothers and sisters did you have? Birth order and spacing.

How old was your father when you were born? (*Prompt*: How old was he when he died? When was that?) Occupation (*If employer*: How many people did he employ?) Did he have another job before or after he became that? Did he also do any casual or part-time jobs? *Continue for all jobs until death.*

Do you remember your father ever being out of work?

How old was your mother when you were born? (*Prompt*: How old was she when she died? When was that?) Had she any jobs before she married? (*If employer*: How many people did she employ?) Did she work after she was married or not? Part-time jobs? Hours. *Continue for all jobs until death.*

Who looked after the children while your mother was at work?

Do you remember your grandparents? Contact, impressions.

*If informants had a substitute parent (e.g. grandparent), adapt questions to include throughout. For stepfamilies, ask about* both *stepparents and natural parents living elsewhere.*

## 2. Domestic Routine

I should like to ask you now about life at home when you were a child; the time when you left school. Can you describe the house at . . . (*Select *from 1*)? How were the rooms used? Bedrooms; other rooms; furniture. Did anyone else besides your parents and brothers and sisters live in the house? Other relatives, or lodgers? (*If lodgers*: Where did they sleep? Where did they eat? How much did they pay?) Did your mother pay anyone to help in the house? (*If yes*: Number of servants; living in, daily or irregular, hours and wages; servants' hall and bedrooms; tasks.) Cleaning; looking after children (time spent by children with parents). Supervision and moral guidance of servants. (*If living in, adapt subsequent questions to establish part in all household activities and relationships*.) How did you get on with her? How did the housework go? Was the washing sent out? Who made or mended the family's clothes? Were any clothes bought new or secondhand? Where and when? Shoes. Did your father help your mother with any of the jobs in the house? *Prompt*: Cleaning; cooking; washing up; fires; decorating; repairs; improvements to the house. Did he dress; undress; bath you; read to you; tell you stories; take you out without your mother; look after you when she was out? Did you have any tasks you had to carry out regularly at home to help your mother and father? How long did you continue to do these tasks? After you left school? *Repeat for brothers and sisters*.

Were you expected to go to bed at a certain time in your school holidays? Did your mother or anyone else put you to bed? Did you share the bed with anyone? Who else slept in your bedroom? Sleeping arrangements of whole family. How did the family manage with washing and bathing?

## 3. Meals

Where did the family have their meals? Were there any occasions when they ate in another room? Who did the cooking? Where? Cooking equipment (range, gas, or electric). When was breakfast eaten? What members of the family were present? How did the others manage for their first meal? What did you usually eat and drink? Did you have anything different on certain days (Sundays)? *Repeat for midday and evening meals*. Did your mother or father bake bread; make jam; bottle fruit or vegetables; make pickles, wine, beer, or any medicines for the family? Did your father or mother grow vegetables and fruit? Did they buy any? (Tinned or dried?) Did they keep any livestock for family (hens, pigs, goats)? Who looked after them? How many times a week did you eat meat? Tinned meat? Did you ever get some extra meat such as rabbit from poaching? Who from? How often? Do you remember seeing your mother having less food so that the family could have more? Did your father have larger helpings? Or extra food?

Were you allowed to talk during meals or not? What was your parents' attitude if you left some food uneaten on the plate? Were you expected to hold your knife and fork in a certain way and sit in a certain way? When could you leave the table? Did all the family sit at the table for the meal? How was the meal served (by whom)? *If employed servants*: Where did the servants eat? Did they have different food?

## 4. General Relationships with Parents: Influence and Discipline

Was your mother an easy person to talk to? Did she show affection? If you had any worries could you share them with her or not? *Repeat for father*.

How did your parents expect you to behave towards them? As a child, was there any older person you felt more comfortable with than your parents? (Grandparents, other relations, servants.) When grown-ups were talking, were you allowed to join in?

What kind of people do you think your parents hoped you would grow up to be? Did your parents bring you up to consider certain things important in life?

If you did something that your parents disapproved of, what would happen? (For example, swearing.) *If punished*: By whom; How; How often; Ever by other parent? Do you remember any particular occasion when you were punished? How did you feel about that?

Would you say that you received the ideas you had about how to behave from both your parents, or did one play a more important part than the other?

How did you get on with your brothers and sisters? Was there one you felt particularly close to? *If quarrelled*: what did your parents say about that?

## 5. Family Activities

When you had a birthday would it be different from any other day? Presents; anything special to eat; guests.

Did you have any musical instruments in the home? Players. Was there anyone in the family who sang? Did you ever make music together as a family?

Did your parents play any games with you? Christmas Day; Easter; other festivals. Were there books in the house? Did you belong to the library? Newspapers. Magazines.

Do you remember a funeral in the family? What happened? Who attended? Did you take part? Did you wear mourning?

Do you remember a wedding in the family? What happened? Who attended?

Were you taken out visiting neighbours, friends, or relations? With whom?

Were you taken shopping? With whom? Do you remember any other outings with your parents? Bicycles; motorbike; car. Bank holidays.

Did you ever go away for a holiday? For how long? Did you do this regularly? Which members of the family went? Where? Activities.

## 6. Religion

Could you tell me how you spent Saturdays in those days? How about Sundays? Did you have different clothes? Did you play games? Did your parents think it wrong to work or play on Sunday? Did your parents attend a place of worship or not? Denomination. How often? Both mother and father? Did either hold any position in the church/chapel? Did you attend?

Did you go to a Sunday School or not? Outings. Choir. Temperance Club Band of Hope. Evening classes. Other activities organized by the church/chapel.

Were you taught to say prayers at night? Did you ever have family prayers?

How much would you say religion meant to you as a child?

## 7. Politics

Did your father take an interest in politics? Do you know what his views were? Why do you think he held those views?

Do you remember your father voting in a general election? Do you know what party he voted for? Did he ever belong to a political party? Activities. *Repeat for mother.*

In some places at that time men felt they risked losing their job or their house if they voted differently from their employers. Do you know if your father felt himself under that kind of pressure to vote for a particular party?

## 8. Parents' Other Interests

When your parents were not doing their work, how did they spend their time? Did they have a radio; TV; record player?

Did your mother have any interests outside the home?

When she went out what did she do? Did she ever go out to enjoy herself? Who did she go with?

When did your father get home from work in the evenings? How many evenings a week would be spent at home? How much was he about the house at weekends? How would he spend the time?

Did your father attend any clubs or pubs? When? Did your mother go too?

Did your father take part in any sport? Did he watch sport? Did he attend the races? Did he bet? Did your mother take part in any sports or games?

## 9. Childhood Leisure

As a child, who did you play with? Brothers; sisters; neighbours. Did you have your own special group of friends? Did you play games against other groups? Where?

What games did you play? Were you allowed to get dirty when you played? Did boys and girls play the same games?

Were you free to play with anyone you pleased? Did your parents discourage you from playing with certain children? (*If yes*: Why?) What did they think about children fighting or gambling in the street?

I should now like to ask about how you spent your free time when you were at school. Did you have any hobbies then? (Collecting—cigarette cards, etc.) Did you keep any pets? Gardening. Did you go fishing? Walks; bicycling. With whom? Dancing or music lessons.

Did you take part in any sports? Did you follow a team?

Did you belong to any youth organizations (Scouts, Guides)? Activities. Theatres; concerts; music halls; cinemas. Did your parents give you any pocket money? What did you spend the money on?

## 10. Community and Social Class

Did anyone outside the home help your mother look after her house or family? (Relations; friends; neighbours.) In what ways? How often?

If your mother was ill or confined to bed how did she manage? Do you remember what happened when one of your younger brothers/sisters was born?

What relations of your father do you remember? Did any live near by? When did you see them? Where? Do you remember them influencing you in any way, teaching you anything? *Repeat for mother.*

Did your parents have friends? Where did they live? Where did they see them? Did they share the same friends? Did your mother have friends of her own? Where did she see them? Did she visit anyone who was not a relation? *Repeat for father.*

Were people ever invited into the home? How often? Who were they? Would they be offered anything to eat or drink? On any particular days or occasions? Would you say that the people invited in were your mother's friends or your father's friends or both of them?

Did people call in casually without an invitation? When?

People often tell us that in those days they made their own amusements. What do you think your parents did when they got together with their friends/neighbours? (Music. Games.)

Many people divide society into different social classes or groups. In that time did you think of some people belonging to one and some to another? Could you tell me what the different ones were?

What class/group (*informant's own term*) would you say you belonged to yourself? What sort of people belonged to the same class/group as yourself? What sort of people belonged to the other classes/groups you have mentioned?

Can you remember being brought up to treat people of one sort differently from people of another? Were you ever told to show respect in some way? To whom? Was there anyone you called 'sir' or 'master'/'madam'? Do you remember anyone showing respect to your parents in these ways?

In the district/village, who were considered the most important people? Did you come into contact with them? Why were they considered important? *If respondent middle or upper class*: Would these people have been considered at that time to be 'in society'?

Where you lived, did all the people in the working (*or* lower *or other term used by informants*) class have the same standard of living, or would you say there were different groups? Describe a family within each group. Do you think that one group felt itself superior to the rest? Were some families thought of as rough, and others as respectable? Do you remember a distinction of this kind between craftsmen and labourers?

Racial groups, immigrants, and religious minorities (clubs, bars, and churches).

Do you think your mother thought of herself as a member of a class? (*Prompt*: middle class, working class?) Why?/Why not? What made her put herself in that class? (*Prompt*: own home background, her job, her type of house, your father's position.)

Do you remember anyone being described as a 'real gentleman'/'real lady'? Why do you think that was? Was it possible at that time to move from one class to another? Can you remember anyone who did?

Was your home rented? *If yes*: What do you remember of the landlord? Did *your* mother or father belong to any savings clubs? Do you know what arrangements your parents had about money? Who paid the bills; made the big decisions? Did they have a bank account; investments? What kind of ideas about money did they give you?

Do you remember feeling that your parents had to struggle to make ends meet? *If no*: Did they help poorer people in any way? Did they belong to any philanthropic organizations? *If yes*: What did you think about that? What difference did it make to the family when your father was ill or out of work? How often. Did you ever get help from the Guardians or the parish or any charity? How did they treat you? How did you feel about that?

## 11. School

Were you given lessons by anyone before going to school?

How old were you when you first went to school?

Type of school (board/private/church; day/boarding; boys/girls/mixed)?

What did you think of school? How did you feel about the teachers?

If you did something the teachers disapproved of, what would happen?

Did the teachers emphasize certain things as important in life? Manners, how to treat the opposite sex; tidiness; punctuality; ways of speaking.

Did they encourage discussion? Was any science taught? Games.

Did your parents encourage you to do school work?

What sort of homes did most of the other children come from? Were some worse dressed than others?

Were there any gangs or groups in the school?

Did you go on to another school afterwards? *If yes: repeat. If a secondary school*: cadet corps; prefects.

How old were you when you left school? Would you have stayed longer if you had had the opportunity? Did you attend any part-time education afterwards? (e.g. evening classes).

*If at university*: Subjects; new friends; new attitudes; influence of tutors; intellectual discussion; religion; clubs and societies; other leisure. How were women regarded at university at that time?

## 12. Work

While you were at school, did you have a part-time job or any means of earning a little regular money? *If yes*: How did you get it? (Through parents?) What exactly did you have to do in this job? How did you learn? Were any practical jokes played on you? What hours did you work? (Saturday; Sunday; half-day). Were there any breaks for meals?

Did you have any holidays with pay? What were you paid? Did you feel that was a fair wage, or not? (Did you give any of the money to your mother? What was it spent on?)

How did you get on with the other people you worked with? Did men and women work together? Could you talk or relax at all? (Could you play games in the breaks?) Was there a works club? A works outing? Any other entertainments for employees? Was there a presentation when a worker retired? Did any of the employers or wives visit workers and their wives at times of sickness or bereavement?

How did your employer treat you? How did you feel about him?

How did you feel about work? Did you like or dislike it? How long did you do it for? When did you give it up? What did you do after that? *Repeat for any other part-time jobs while at school.*

Now I should like to ask you about your first full-time job. What was that? *Repeat questions above, for all jobs (including part-time) to retirement. These questions are schematic and much fuller questions and promptings are desirable for main occupations.*

Did you serve an apprenticeship or training period for any of your jobs? Did you (or any of your employees) belong to any trade union/ professional organization? Did you take part in any of its activities? Did you feel that employers had the same interests, or different? Did you feel that there were divisions of interest among workers?

Were there any chances for promotion? Would you have preferred another type of occupation yourself?

*If an employer or manager*: Can you tell me who owned the business? How was it founded? How was it run? How did you learn about the different sides of the business (technology, sales, staffing, finance)? Which interested you most? Did you become a partner? What share did you have in the profits and losses? Did senior partners/directors share a social life together? What did the workers call you? Did you meet any of them outside work?

## 13. Home Life after starting Full-time Work/leaving School

I'd like to ask you about your life at home after you started full-time work (or left school). Did you continue to live at home then? For how long? *If at home*: Did you have our own room where you could entertain friends privately? *If separately*: Did you live alone or share with anyone? Describe house. Did you have any domestic help? Where did you mainly eat?

*If working*: Did starting full-time work change your relationship with your parents at all? How much of your wage did you give to them?

*If not working*: How did you manage for money? Would you have rather done something else? How did you spend your time (housework, social calls, family business)? Did you spend your Sunday any differently? (Church/chapel; Sunday school.) Did religion mean more or less to you after childhood? Did you start to take an interest in politics? Or later? Activities. Can you tell me something of how you spent your spare time as a young man/woman? Did your interests change? Clubs or youth organizations; sports or games (cards; tennis); dances; hobbies; outings; theatre, music hall, cinema; pubs? Did you go out in the evening? Where to? Who with? What was a good night out in those days? Holidays. Where; who with? Did you make any new friends—boys or girls—at this time? How did you meet them? Did you stick to a group of friends? Workmates; shopping friends. What did you do with them, talk about? Did you have any special friends at this time? Boys or girls? Were there any special places where boys and girls could meet? Where would you go with them? Were you allowed to be with them alone? Did your parents meet your friends? Did they expect to know where you were? Did you have to be home by a certain time? Did your parents disapprove of any of your activities at this time? Smoking; sex. Do you remember your parents' attitudes towards sex?

## 14. Marriage

What age were you when you married? How long had you known your husband/wife then? How did you meet? Where did he/she come from? What kind of family? How long were you engaged? Did you save up money before getting married, or not? Did your parents help you in setting up a home? Did they help you later on? (or leave you anything?) (Or by that stage, did you have to help them?) Could you describe the wedding? Did you have a honeymoon? Where did you live after you married? How many years? (Did you ever consider moving out of the area when you first married?) Where did you live then? *Continue for subsequent moves.*
How old was your husband/wife when you married?
*If woman*: What was your husband's job when you married? Did he have other jobs before or after? *Ask for all jobs*. Did he also do any casual or part-time jobs? *If informant worked after marriage*: How did your husband feel about your working? *If man*: Did your wife have a job when you married? Had she any other jobs before that? Did she continue working after your marriage? *If yes*: How did you feel about that? What jobs had she had since then? *Ask for all jobs*.

How far did you get your ideas of a good marriage from your parents?
How important would you say sex is to a marriage relationship?

## 15. Children

Did you have any children? How many? Names and years of birth. Did
you plan to have the number you did? Contraception.
Were your children born at home?
*If woman*: Did you know what to expect in childbirth? Books; classes.
How did you get on? Who else was there? How did you feed your first
baby? Did you have any difficulties in feeding? If you needed advice,
who did you ask? Did you punish it when it was naughty? How? For
what? How much did your husband have to do with your children when
they were babies under one year? Did he do more for them later on?

## 16. Family Life after Marriage

*Budget and control of household*: I want to ask you how you and your
husband/wife managed the housekeeping in those years.
*If husband*: How much of your earnings would you give to your wife at
that time? Did you pay any of the bills yourself?
*If wife*: Did you know what your husband earned? How much of that
would he give to you? Did he pay any of the bills himself? How did you
decide the money should be spent? (Who chose new furniture; food;
drink; doctor; church; clothes of children, husband; presents; outings;
holidays; who should be invited to stay or to meals? Who looked after
the garden?)
What did you do when you disagreed? How would you describe the
relationship you had then? Did you talk to each other and share important things?
*Return to section 2, and repeat with appropriately modified phrasing
through to section 11; in section 4 add*: When your children were young
did you feel that there was a right way/wrong way of bringing up children? Did you and your wife/husband have the same ideas about
bringing up children, or different ideas? Did you talk about this? Was
there anyone you used to talk to if you were worried about the children?
Was your mother alive when your children were small? How often did
you see her? Did you ask her advice in bringing up the children? Mother-
in-law.
Did you believe that girls should be treated the same way as boys when
you had your children? That they should be taught the same skills and
the same games (e.g. girls carpentry, hunting; boys sewing, cooking,

dancing, piano)? How did you teach your boy to behave to his sister (e.g. opening doors, carry things)?; your girl to her brother (sew for him, wait on him)?

*If wife worked after having children*: Who looked after the children while you/your wife was at work? How did you feel about leaving the children with somebody else? Some people think that children should be with the mother all the time, others think it is not necessary and does them good to be with other people quite a lot too. What did you think at that time?

*Unless the informant has moved to another community since childhood, do not repeat the mid-section 10 sequence on social class. End the interview by asking about children's schooling and subsequent occupations.*

# Index

## MORE OXFORD PAPERBACKS

Details of a selection of other books follow. A complete list of Oxford Paperbacks, including The World's Classics, Twentieth-Century Classics, OPUS, Past Masters, Oxford Authors, Oxford Shakespeare, and Oxford Paperback Reference, is available in the UK from the General Publicity Department, Oxford University Press (JN), Walton Street, Oxford OX2 6DP.

In the USA, complete lists are available from the Paperbacks Marketing Manager, Oxford University Press, 200 Madison Avenue, New York, NY 10016.

Oxford Paperbacks are available from all good bookshops. In case of difficulty, customers in the UK can order direct from Oxford University Press Bookshop, 116 High Street, Oxford, Freepost, OX1 4BR, enclosing full payment. Please add 10 per cent of published price for postage and packing.

## A WORLD APART

*Gustav Herling*

Gustav Herling's remarkable account of his experiences in a Soviet labour camp during the Second World War was instantly hailed as a classic. He relates the horrors of life at Kargopol camp, where he was imprisoned for one and a half years. It is without doubt one of the most moving and important books in the literature of oppression.

## AN OLD WOMAN'S REFLECTIONS

*Peig Sayers*

*Translated by Séamus Ennis, with an introduction by W. R. Rodgers*

Peig Sayers, 'the Queen of Gaelic story-tellers', spent the greater part of her long life on the Great Blasket Island. She was a natural orator, and students and scholars of the Irish language came from far and wide to visit her. In this book, as an old lady, she muses and reflects on the days of her youth, recounting tales which evoke characters and an era now dead, and capture the superstitions and hard life of her beloved island.

## LIFTING THE LATCH

### A Life on the Land

*Sheila Stewart*

For nearly eighty years Mont Abbot has lived and worked on the land near Enstone in Oxfordshire. This extraordinary record of his life and times was constructed by Sheila Stewart from a series of taped conversations.

## MY STORY

*Peter O'Leary*

Praised and damned in equal measure since it was first published in 1915, 'My Own Story' (*Mo Scéal Fein*) recounts in a forthright and unrepentant tone the life of Father Peter O'Leary, priest and teacher, cultural and social reformer, Gaelic revival activist, author, and controversialist. His long life covered the key events of modern Irish history—the Great Hunger, the 1848 Rebellion, Tenant Rights, Fenian '67, the Land War, Home Rule agitations, the Gaelic League, the Easter Rising, and the War of Independence—and in many cases he was not just a witness but an active participant.

This translation, the first into English, has been made by Cyril O'Céirín, who has also added notes and appendices to the text, and provided a new introduction.

*Forthcoming*


## A PITY YOUTH DOES NOT LAST

*Micheál O'Guiheen*

*Translated by Tim Enright*

This is the only English translation available of *A Pity Youth Does Not Last* by Micheál O'Guiheen, son of Peig Sayers, and the last of the Great Blasket poets and story-tellers. In it O'Guiheen writes of a childhood spent on the Great Blasket, and of the changes that finally overtook the old island culture.

'A sweetly elegiac memoir of life on Great Blasket Island . . . He wrote it in Gaelic and it has been translated into "Kerry English" by Tim Enright, a translation that has the right raw sound; that captures the drift of a story that is ancient and as timely as any to be met with in the Nordic Saga.' *Country Life*

## THE JOURNAL OF A COUNTRY PARISH
### Robin Page

From his Cambridgeshire farm the author shows us the realities of life in his village and the surrounding fields through one year. 'Robin Page has captured all the flavour and charm of his own countryside.' Gordon Beningfield.

## ISLAND CROSS-TALK
### Tomás Ó'Crohan
### Translated by Tim Enright

In these pages from his diary, Ó'Crohan jotted down snatches of conversation, anecdotes, descriptions of the landscape and the sea.

*Island Cross-Talk*, first published in 1928, was the first book to come out of the Blasket Islands, that remote, tiny community off the West Kerry coast speaking a dying language. It sowed the seeds of a rich and extraordinary flowering of literature: Maurice O'Sullivan's *Twenty Years A-Growing*, Peig Sayers's *An Old Woman's Reflections*, Ó'Crohan's later book, *The Islandman*, and many others.

## OUR VILLAGE
### Mary Russell Mitford

The little village of Three Mile Cross in Berkshire was Mary Russell Mitford's home for thirty years. She has drawn on her observations of the locality for many of her short essays, the best of which appear in this book, to give a unique picture of country life in the early years of the nineteenth century. Village events and festivals are described in vivid detail, as are the many colourful characters who peopled the small neighbourhood.

## DEAD AS DOORNAILS

A Memoir

*Anthony Cronin*

The seven men portrayed in this book—among them Patrick Kavanagh, Brendon Behan, and Myles na Gopaleen (Flann O'Brien)—were brilliant literary figures in the artistic world of the fifties and sixties. They led extrovert and eccentric lives at a time when conformism was still the order of the day. The poet and broadcaster Anthony Cronin knew them all, and in this memoir analyses them with a sympathy and honesty.


## A YEOMAN FARMER'S SON

A Leicestershire Childhood

*H. St G. Cramp*

The author describes his childhood in the village of Tur Langton in Leicestershire in the 1920s. 'A marvellous account of a family, a farm, and a village, written with verve and gusto.' *Dorset Life*


## THE ISLANDMAN

*Tomás Ó 'Crohan*

*Translated from the Irish by Robin Flower*

Tomás Ò'Crohan was born on the Great Blasket Island in 1856 and died there in 1937, a great master of his native Irish. He shared to the full the perilous life of a primitive community, yet possessed a shrewd and humorous detachment that enabled him to observe and describe his world. His book is a valuable description of a now vanished way of life; his sole purpose in writing it was in his own words, 'to set down the character of the people about me so that some record of us might live after us, for the like of us will never be again'.

# PART OF MY LIFE

## A. J. Ayer

In this first instalment of his witty and candid autobiography, A. J. Ayer revives memories of his family background, his experiences at Eton in the 1920s, as an undergraduate at Christ Church, Oxford, as a student in Vienna, and as a young don in the 1930s. He traces the growth of his interest in philosophy, and his philosophical development under the influence of Gilbert Ryle, Wittgenstein, and the Vienna Circle. The book tells the story of the author's first love and marriage, and contains portraits of many of his friends, including Bertrand Russell, E. E. Cummings, George Orwell, Goronwy Rees, and Isaiah Berlin.

'It is the unflinching truth-telling which gives the book its importance.' C. P. Snow in the *Financial Times*

'It is a sensual as well as an intellectual biography . . . It is a very honest book, full of fascinating vignettes from a worldly, exuberant life . . . a pure pleasure to read.' Tom Stoppard

# WITH O'LEARY IN THE GRAVE

## Kevin Fitzgerald

There are two stars of Kevin Fitzgerald's hilarious, yet poignant, autobiography of his early years. One is his preposterous, infuriating, but charismatic father, whose complex business affairs and extraordinary whims were his family's despair. The other is rural Ireland of seventy years ago, where the land was ploughed by horses and no worthwhile dance ended before daybreak. Fitzgerald tells his story (which also takes in London and the Canadian prairies) with great humour and eloquence, and plenty of just-believable anecdotes.

'There are few autobiographies of which one can say that they are too brief and that there ought to be a sequel, but this is one.' Anne Haverty in the *Times Literary Supplement*

# THE JOURNAL OF A SOMERSET RECTOR, 1803–1834

*John Skinner*

*With an essay by Virginia Woolf*

*Edited by Howard and Peter Coombs*

John Skinner's journal reveals many truths about life in rural England at the beginning of the nineteenth century. He spares us no detail of the appalling social conditions and injustices he found in his small country parish. Virginia Woolf's brilliant essay hints at a special affinity she felt with this neurotic, introspective man, who, as she herself was to do, fell victim to a final suicidal depression.

'An extraordinary document, historically and personally.' *Country Life*

'a great find . . . fascinating' *Open History*

'a fascinating insight into the poverty and suffering which was English rural life during and after the Napoleonic wars' *Tribune*

# THE GATES OF MEMORY

*Geoffrey Keynes*

Geoffrey Keynes had, as he put it himself, 'a quite outrageously enjoyable existence'. This is his remarkable account of a long and distinguished life, written only two years before his death in 1982.

The younger brother of the economist Maynard Keynes, Geoffrey Keynes was at Rugby and Cambridge with Rupert Brooke. He saved Virginia Woolf from her first suicide attempt, became a celebrated surgeon who pioneered the use of blood transfusion and the rational treatment of breast cancer, and was knighted for this work in 1955.

'less an autobiography than a valuable piece of social history . . . a portrait of an age' Anthony Storr in the *Sunday Times*

## BLUE REMEMBERED HILLS

### A Recollection

*Rosemary Sutcliff*

Rosemary Sutcliff is one of our most widely acclaimed novelists for children (and she has many adult admirers too). In *Blue Remembered Hills* she gives a moving account of the influences and the people that helped in her personal development as a writer.

'It is a remarkable book, not only for the clarity of her memory and for her determination to be honest, however painful the revelation, but also for her considerable powers of description' Caroline Moorehead in *The Times*

## MORE OF MY LIFE

### A. J. Ayer

This sequel to *Part of My Life* is A. J. Ayer's autobiography of his middle years. He writes with instinctive modesty and honesty, finding humour in his failures and qualifications to his successes.

'This is an engaging funny book that will live to be quoted.' George Watson in the *Financial Times*

## THE WESTERN ISLAND

### Robin Flower

Dr. Flower spent a considerable amount of time between 1910 and 1930 living amongst the 150 inhabitants of the Great Blasket Island. He tells of the adversities and frugality of the Gaelic-speaking people, of the folk-tales and the stories of ghosts and fairies and poets.